J. A. of Lambertville Anderson

The Train Wire

J. A. of Lambertville Anderson

The Train Wire

ISBN/EAN: 9783743372887

Manufactured in Europe, USA, Canada, Australia, Japa

Cover: Foto ©berggeist007 / pixelio.de

Manufactured and distributed by brebook publishing software (www.brebook.com)

J. A. of Lambertville Anderson

The Train Wire

THE TRAIN WIRE

A DISCUSSION OF THE SCIENCE

OF

TRAIN DISPATCHING

BY

J. A. ANDERSON;

WITH AN INTRODUCTION BY B. B. ADAMS, JR.

Second Edition—Revised and Enlarged.

PUBLISHED BY

THE RAILROAD GAZETTE, 73 BROADWAY, NEW YORK.

1891.

74885

CONTENTS.

INTRODUCTION.

In the first edition of this book, issued in 1883, Mr. Anderson, then Superintendent of the Belvidere Division of the Pennsylvania Railroad, modestly disclaiming perfection for his work, ventured the prediction that the science of which he wrote would be greatly advanced as time went on. In one sense this prediction has not been fulfilled. The eight years which have passed have witnessed little or no change from the principles then laid down by the author of The Train Wire, but he has the satisfaction of now seeing their widespread adoption and a consequent great improvement in the practice of this important science; and while probably none at the present time know how to handle trains by telegraph better than the dispatchers of the Pennsylvania road did when the prediction referred to was made, the requisite knowledge and training are now possessed by many more men than were numbered among the experts of the earlier period.

The author's disavowal of exhaustive treatment is proper in view of the fact that a complete treatise on the subject would include much relating to the operation of the train rules and to points of discipline ; but it must be agreed that the first edition of this book was the first thorough and precise essay on the subject which had appeared, and that it stated the principles of dispatching in substantially the form since adopted by the General Time Convention, a body composed of the General Managers and Superintendents of practically all the important roads of the country east of the Missouri River.

The inception of this book resulted from the author's work, several years earlier, in revising the rules of the company under whom he was employed; and in preparing his book he naturally took care not to trespass upon the prerogatives of that company; but it is no more than right to say that outside observers regard his work as one for which his own road and all others are as much indebted to him as he can be to any road.

During the preparation of the Rules on Train Dispatching, formulated by the eminent Managers and Superintendents composing the Time Convention Committee, Mr. Anderson acted with that Committee, and his suggestions in The Train Wire, with his other work in that line, were largely used as the basis for this portion of the Time Convention rules. The deviations in these rules from the lines laid down in the first issue of The Train Wire are chiefly in the nature of compromises as to methods of practice, made necessary to effect an agreement among railroad officers of different needs and opinions. The Standard Code avowedly falls short of perfection, but chiefly because of this necessity.

The duplicate form of order is presented by Mr. Anderson as a vital feature in the science of dispatching. When he first wrote, this form of order was in use on few roads. Many officers were ignorant of it, and most others knew of it only in a vague way or looked upon it with disfavor as impracticable for roads doing a heavy business. Now, the requirement that all trains concerned in the execution of a specific movement should receive the order in the same words, is widely recognized as an axiom, and rules based on this principle are fast coming into general use.

The first part of the book treats of general principles, while the latter part takes up the rules which embody those principles and give them effect, the Standard Code being taken as the basis of the discussion. It might at first seem unnecessary, in view of the wide acceptance of the Standard Code, to enter into a discussion of its rules, and some of this discussion may appear to be needless repetition of matter presented in earlier pages; but as there are still those who have not taken the most advanced position, and probably many who, having adopted good practice, are not thoroughly familiar with the reasons for it, the author has done well to retain this feature of his earlier work, in connection with the statement of principles. These comments serve to point out to those not thoroughly acquainted with the subject the relations of the rules to the reasons for them, and this must be useful to beginners in the science and to men on new roads. For officers of experience, whose positions remove them from personal contact with the telegraph work and yet require that they have particular knowledge of it, a book of this kind should be both elementary and full; and all readers will find in examining the rules for practice that there is an advantage in having attention directed to the conformity of the rules with the principles before enunciated.

One of the most interesting and original paragraphs in the first edition of The Train Wire was that describing the scheme for numbering switches and using those numbers in train orders, to facilitate the movement of trains at meeting points. This plan has since been put in use to some extent and has given great satisfaction; and in connection with "lap-sidings" it has been found of marked benefit in handling a heavy traffic on a single-track road.*

The author of The Train Wire is no longer connected with the Operating Department,† and has undertaken this revision reluctantly, but his interest in his former work

---

*A description of the use of lap-sidings and numbered switches on the Cleveland & Pittsburgh Railroad was published in the *Railroad Gazette* of December 26, 1890.

† He is Superintendent of the Voluntary Relief Department of the Pennsylvania and allied roads, with office at Trenton, N. J.

is still lively, and this is an enlargement as well as a revision ; so that both reader and author are to be congratulated. The superintendents and dispatchers, the operators young and old, among the million railroaders of the United States, have a better handbook than ever before, while the author can justly take pride in the fact that the individual views expressed by him in 1883 are now generally accepted truth. The introduction of the Standard Code on 70,000 miles of American railroads is one of the important steps of recent years in railroad operation, resulting in greater security to life and property ; and The Train Wire should be credited with a liberal share of the honor of the reform.

B. B. A., Jr.

PREFACE TO THE FIRST EDITION.

The views on train dispatching here offered have been arrived at during an experience of some twenty years, including a recent connection with the preparation of a set of rules for the company on whose road the writer is employed. While his agency in the formation of the rules referred to accounts for the existence of a general similarity and no radical difference between them and the present treatment of the subject, the latter is not to be taken as an authorized commentary upon those rules, but as an expression of individual views for which, with any additional matter or variations in arrangement, the writer is alone responsible.

With his first experiments in train dispatching the writer became convinced that the method of issuing train orders in the same words to all concerned in each transaction afforded greater security than that supplied by any other form of order. Another early conviction was that each step in the process of preparing and issuing train orders should be carefully and minutely arranged for by specific rules.

In here undertaking to impress these views, it is also sought to set forth the general principles upon which rules should be based, and to recommend methods of procedure for all ordinary practice. The methods proposed have been tested by the writer, and the most of them by others. If they are not found to apply to all existing circumstances, they may at least serve as guides in devising other plans.

It is not assumed that this consideration of the subject of train dispatching is exhaustive. The theme is a fruitful

one and of growing interest and importance. Much remains to be said of what has already been accomplished,
and the future will doubtless show advances in this science
far beyond the best practice of the present.

1883.

THE TRAIN WIRE.

CHAPTER I.

TRAIN DISPATCHING.

The telegraph, as a means of directing the movements of trains, is a necessary railroad fixture. But for its agency the moving of the heavy traffic of some of our railroads would be impossible without large additions to the tracks and consequent increase in the cost of construction and transportation.

The train wire is thus a promoter of both economy and facility of operation. Under the supervision which it permits, the products of industry are rapidly and cheaply exchanged between distant points, while the traveler, unimpeded by the slower-moving trains, goes swiftly on his way. Steam is the noisy giant that shoulders the load and gets the praise ; but the silent man, in some quiet place away from the rattle of the wheels, with his finger on the key, controls the ponderous and complicated movements, which proceed so harmoniously that one may almost imagine them to be the result of natural law.

Although the value, however, of the telegraph as a railroad appliance is daily becoming more fully realized, its capabilities for usefulness have not been developed to an extent commensurate with its importance. A well-informed writer has justly said : " Telegraphy as a handmaid of the railroad has not assumed any enduring form peculiarly adapted to this business."

This is still true in a measure, although not to so great an extent as when uttered. The circumstances must be very exceptional in which the aid of the telegraph will not be of important advantage. Machinery breaks, steam fails, connections are late, storms and floods disturb the roadway ; a thousand things cause delays. The difficulties may not be great or numerous where trains are few, but they increase rapidly with the growth of traffic, and vexatious delays can only be avoided by adequate means of promptly controlling the movements of the trains. Hence the importance of securing not only the best telegraphic appliances, but the best method as well of rendering them useful in the service in question.

In arranging a system of train dispatching, its relations to safety and economy require that careful consideration be given to the principles on which it should be based. Some of the methods in use indicate this careful study and a growing sense of its importance is shown in the recent general acceptance of rules on the subject, prepared with the most scrupulous care.

These rules, as will be seen, are in conformity with what was urged in the former edition of this work, and the present intention is to direct attention anew to some of the underlying principles, as well as to the practical bearing of the rules referred to.

The means of instant communication afforded by the introduction of the telegraph seemed to place at command a method of directing distant train movements with ready facility ; but it soon appeared that the use of the new implement involved risks which must be carefully guarded against ; hence the various " systems" which have arisen having this in view.

The distinctive feature of the "American" system of train dispatching is the issuing of orders from a central office, directing train movements, supplementary to those provided for by the time-table and "train" rules. This method is in general use, and is recognized as better adapted to our circumstances than that of moving trains by the "staff" or other means from station to station, as in European practice. In considering the application of this mode of issuing telegraphic orders for single-track, some of the methods will be seen to apply as well to roads having more than one track.

A printed time-table, showing the regular times and meeting-places of trains, may be prepared at leisure and studied by all train-men, and is full notice as to all regular trains on the road. With rules added directing how the trains are to proceed with relation to each other, understood by all alike and faithfully observed, collisions cannot occur. If, however, it becomes necessary to issue special orders for trains that are not on the time-table, or for the forwarding of any, otherwise than by the opera-tion of the ordinary rules, new precautions be-come necessary.

The conductor or engineman receiving such an order must know *that it is given by competent authority.*

It must be understood *that others concerned have corresponding orders.*

These orders should be *so clearly expressed that they cannot be misunderstood,* and they should be forwarded and delivered *under such safeguards as to insure their certain and cor-rect reception by the proper persons.*

As these orders are to be acted upon at once, without opportunity for careful study, *their form, and even the paper on which they are written, should be such that they may be easily and quickly read and comprehended.*

It is now generally agreed that *orders of this kind should be issued by a designated dispatcher*, acting by the authority and in the name of the superintendent. For two persons to engage in this work at the same time for the same piece of road involves serious risk, and to insure safety as well as confidence on the part of the trainmen this should never occur. It may be taken as an initial principle that *the success of a system depends largely upon the assurance upon the part of the trainmen that every source of danger has been carefully considered and guarded against, and that the rules adopted are strictly adhered to.* If it were known, for instance, that orders were issued by the superintendent and one of his assistants alternately, as might be convenient at the moment, it would excite distrust. The author must confess to such feeling when, some years since while on a delayed passenger train at a way station, he saw the superintendent take a bit of paper from his pocket and write against the side of a building an order for the train to proceed to a certain point, regardless of another designated train. It came out all right, but the incident did not inspire confidence in the telegraphic system of that road. Within the knowledge of the author a disastrous collision resulted from an oversight in regard to the delivery of an order where a skilful official undertook to assist a dispatcher in an emergency. Between the two an important point was omitted ; each thought the other had attended to it. Extreme care is necessary to carry out exactly the methods fixed upon for the proper preparation and issuing of these messages, and confusion is likely to result from interference with those charged with this duty.

In issuing a time-table in advance of the date upon which it takes effect, means can readily be used for making sure that it is received by those who are to be governed by it. The means are more complicated and subject to greater risks whereby we can be assured that a telegraphic train order reaches correctly and surely the hands of those for whom it is designed. After preparation by the Dispatcher it is transmitted in telegraphic language by mechanical agency to a distant point, there to be retranslated into plain English and written out without mistake, for record and delivery ; and all this in the shortest possible time.

The details of this process should be so arranged as to guard as far as possible against every risk arising under the several steps, and *nothing should be left to mere personal care that can be provided for by fixed methods of proceeding.* To one who is an expert and can see in his own case no occasion for extraordinary safeguards such precautions may not seem important; but a consideration of the risks involved, of the many steps to be taken, and of the number of agents engaged in the process, many of whom are often not greatly experienced, must lead to the conclusion that *a methodical following out of a carefully prepared mode of proceeding* is a most valuable means of providing against many of the chances of failure.

Two general methods or "systems" of constructing train orders are in use. They have been distinguished as the "single order" and "duplicate order" system. The latter is accurately described by its title. The other title is not a strictly accurate designation, but sufficiently so for our purpose.

Although the "duplicate" method is now widely recognized as the best, the other is still in use. For purposes of comparison of these methods we will take a telegraphic order providing for the meeting of two trains at a designated point beyond which the one has, by train rules, the superior right of track as respects the other. The order is to limit the superior right, and permit the inferior train to run to a point to which it could not otherwise go without trespassing on the right of the other. If by any error or misunderstanding the superior train fails to stop at the proposed meeting-point, while the other proceeds upon the assumption that it will thus stop, the result may be a disastrous collision.

Under the "*single order*" system, when two opposing trains are to meet by special order, arrangements are usually first made to stop the superior train by a "holding order." An order is then given forbidding it to go beyond the designated point, and then another order is given to the inferior train authorizing it to go to that point.   The holding order is addressed to an agent or operator whose station the superior train will pass, and reads substantially as follows :

Hold train No. 5 for orders.

The person receiving this is required to display a signal to stop the expected train if it is not already at the station, and not to allow it to proceed until the meeting order is duly forwarded and delivered.   This order to the superior train is usually addressed to the conductor and engineman in the following form, or its equivalent :

You will not pass Alton until train No. 4 arrives.

The corresponding order to the conductor and engineman of the inferior train, sent to some station to be passed by it, will read :

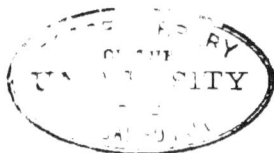

You will run to Alton regardless of train No. 5.

or perhaps—

You will meet and pass train No. 5 at Alton.

The holding order is dispensed with by some, and with some it is the practice to issue orders to inferior trains while a superior is held by a holding order until its movements can be determined on, when it receives an order covering all that have been given to trains against it.

Under the "*duplicate*" system the holding order may be used, but such has not been the general practice, and it would not under this system be used in the manner above described. This system, as its name implies, requires that *the order given to each train shall be a duplicate of that given to every other train* concerned in the movement provided for in the order. For the simple movement above described an order is addressed to the conductor and engineman of each of the two trains, *in the same words,* as follows :

Trains No. 4 and No. 5 will meet at Alton.

This, being in the same words to each, may be transmitted over the wire to both at the same time. This is usually done, and offers one of the chief advantages of this form of order. The trains are stopped by signals, which are required either to be displayed when an order is sent, or to stand normally in position to stop trains, which are only permitted to pass on the signal being changed or on getting proper orders.

Objection has been made to the "duplicate" form that it does not distinctly order a train to proceed farther than its schedule rights permit, nor in definite terms direct the other not to go beyond the new meeting-point. The objection has no weight, as an order to meet can only be construed as authorizing each train to go to the station named, and not beyond it until both are there; and it is easy and proper to provide a rule which shall definitely settle the point for those who are unaccustomed to this form, if it should be deemed necessary.

The fatal defect in the "single order" system is that the orders to the two trains, written separately and differently expressed, are subject to the grave danger of inadvertently giving in one a meeting-place different from that given in the other. This liability is greater if an interval of time occurs between the preparation of the two. The risk is very much increased by the usage under this system of including several meeting-points in one order, and becomes still more serious if meeting-points are to be made for several trains moving in each direction. The schedule for these must be rapidly made up and written out in parts, giving to each train its part, differing in form from all the others. There is nothing but the care and skill of the Dispatcher to prevent the opposing orders from differing in some particular. When we consider the care necessary in preparing a time-table, to properly show the running time and meeting-places of the several trains, we must see that the risk, in the process described, of getting something wrong, must far outweigh any supposed convenience in a train having an order showing a continuous schedule of its meeting-points for several opposing trains. Those unacquainted with this work would be astonished at the extent to which the skill of some dispatchers in this direction has been developed. To the uninitiated the mental operations would be simply bewildering, which are required of a brain from which issue for hours, without apparent effort, the instructions under which the trains on a busy road are moved expeditiously and harmoniously. It is not to be denied that many men have moved traffic of huge dimensions safely and with entire satisfaction by the "single order," but this does not at all prove that the system possesses inherent principles of safety. Great personal ability and skill have, with it, achieved marked success where in less able hands its defects would have become apparent; but that some have developed this remarkable ability is no reason why we should depend upon this

in a matter of such vital importance. The
prevalence of methods which require exceptional skill has doubtless interfered with the
more extended usefulness of the railroad telegraph which would probably have resulted
under a system more readily operated by men
of less experience and ability.

Men who have successfully worked under
the "single order" method have stated that the
mental strain is very great, augmented by anxiety born of the fact that a single error may
be fatal to property or life. Now, a mode of
constructing orders which may be operated
with safety by men of moderate skill, which
relieves them of the mental strain, and *which
in itself provides against the most serious
chance of error* must at once commend itself.
The "duplicate" would appear to meet these
requirements ; and that such is the case is the
abundant testimony of those who have used it.

In preparing this order the Dispatcher cannot possibly give different meeting-points, as
there is but one message for both trains, and
when transmitted to both simultaneously each
must get the same as the other. The mental
anxiety arising from the other method is
absent in this. An experienced Dispatcher
under the "single" system has stated that
in visiting an office where the "duplicate"
was used he was surprised that those engaged

there appeared to have so little on their minds. He found, on himself adopting the "duplicate," that it was readily explained. Each transaction is at once complete. On the preparation and transmission of the order in precisely the same language to both trains, and with no necessary connection with any other transaction, the mind is at once prepared to dismiss that and go on to the next. In the transmission of two separate orders for the one meeting, there is ever the feeling that an error may be or may have been committed. But where the one sentence is prepared for both trains and, as is usually done, transmitted to both at one sending, the Dispatcher may rest secure that *no collision can occur from any oversight of his in preparing the orders*, and superintending officers may, if necessary, commit this work to comparatively unskilled hands, with the assurance that so long as the prescribed methods are adhered to the proceeding will be *at least safe*, however great may be the delays arising from unskilful movements.

The power of combination and of quickly calculating the probable movements of trains and determining what shall be done is an entirely distinct matter. This power is largely the result of experience. It is essential to the full development of any system, but is exer-

cised with much greater facility under the
relief which the "duplicate" affords. It
has. been alleged that this method requires
more telegraphing than the other, and that
trains cannot be moved by it so promptly.
It has, however, been for many years in use
on roads where only the most expeditious
methods would serve; and superintendents
moving a heavy traffic, who have changed from
the "single" to the "duplicate," state that
the amount of telegraphing is reduced. one-
third. Those who have grown up with a system may have reasonable hesitation as to
making a change. It is not easy to give up
methods of practice in which one has been
trained for those which are new; and it may
seem difficult, perhaps unsafe, to undertake to
re-educate operators and trainmen in so critical
a matter. Nevertheless, those who have tried
it have found these supposed difficulties to
quickly vanish, and have discovered the result
to be in every way satisfactory, and that this
form of order is much to be preferred. Some
officers who were with difficulty induced to
change are now among the most enthusiastic
supporters of the "duplicate" method.

In arranging for the issuing of train orders,
experience has shown that forms may be simplified and improved methods adopted by
which the work is facilitated and the orders

rendered clearer to those receiving them ; and disaster has taught the necessity for precautions not before thought of. These points will be considered in detail with reference to the "duplicate" system of orders, although much that follows will apply to the other.

CHAPTER II.

The Train Dispatcher holds a most import- ant position as respects safety of life and property. He may perhaps do more than any other official to secure it by care or endanger it by lack of vigilance. His relations to economy, too, are important. As the time of engines, cars, and employés,and of the persons and things carried, is of value, delay avoided is money saved.

It cannot be too strongly insisted upon that the man who issues train orders should make it his especial business, and should have no interference from others. None but a very limited business will warrant the performance of this duty by the superintendent in person, or by any one engaged in other work. If it is such as to call for any approach to continuous attention, persons must be specially assigned to it. The hours of duty and the question of other occupation must depend upon the frequency and constancy of the demands of the work specially in hand. Upon a busy road where the trains are run much on orders, safety as well as efficiency will be best promoted by

2

excluding other occupation and anything
which may distract the attention of the "Train
Runner," and under these circumstances a
period of duty of eight hours is as long as can
prudently be assigned. This conveniently di-
vides the twenty-four hours between three
men, and does not overtax them. With lighter
duties a longer time may be admissible. With
very heavy work, six hours may be long
enough.

The importance of confining the work of
dispatching, for the time being, to the individual charged with this duty, has already
been referred to, and cannot be too strongly
urged. The office where this work is done
should be separate from others, and should not
be subjected to the visits and conversation of
outsiders or of employés whose business does
not call them there. The Dispatcher should
be a proficient operator. He may not himself
transmit his orders, but he should be able to
read all that passes on the wire, in order
to have an intelligent understanding of
what is going on. He should be thoroughly
acquainted with the location and length of
the various sidings, the grades and curves,
the capacity of the engines, and other matters
which may affect the movements of the trains
he has in hand, and some experience as conductor will be of value. He should be a man of

more than average ability, of good judgment, clear head, and strictly temperate habits. In many cases the chief Dispatcher is the right-hand man of the superintendent in all matters associated with the management of the trains; and a suitable recognition of the importance of the position will have a valuable effect in elevating the character of this service and in promoting its efficiency.

CHAPTER III.

Where the work of the Dispatcher is considerable, he will require the aid of one or more operators in the work connected with the transmission of orders. In view of the importance of his duties and of the fact that he may in turn become Dispatcher, the operator should be selected with care. He too should have a clear head and correct habits, be a good penman, an expert telegrapher and a sound-reader. It will be his duty to transmit the orders, or write them down as transmitted by the Dispatcher, and to follow them through the subsequent steps until the process, up to delivery, is complete. He should not be charged with message or clerical work where it may interfere with his principal duty.

The station operator who receives the orders must also have part in the subsequent steps, and on him is placed the duty of delivery. Besides the personal and professional qualifications required for the other, he should, with him, be thoroughly conversant with the rules and methods prescribed for this service, as well as with the time-tables and general train rules and the character and designations of the trains. A station operator may do much to keep business moving by advising the Dispatcher of arrivals, delays, and other things occurring near him, which have a bearing on train movements, but which the letter of his instructions may not require him to report. One who does this intelligently prepares and recommends himself for advancement.

It is quite important that operators be impressed with the gravity of the work in hand. Their apprenticeship and training should be such as to assure this as far as possible, and before appointment they should be thoroughly examined as to their qualifications in all respects, and afterward constantly supervised by competent officials. Young persons readily learn to telegraph, and the lowest compensation paid is something considerable to the youth just leaving home, while the salaries usually paid to railroad operators are not such as to offer fair inducement to men of years and experience to accept or retain these positions. Hence many of our operators are comparatively young. It is no disparagement to them to say that they have not ordinarily the steadiness of character and sense of responsibility which we expect in maturer years. Without these it is difficult for them to have a proper conception of the magnitude of the interests dependent on their attention to their duties, and of the importance of exactly carrying out details which to them may seem almost trivial. We have here a cogent reason for so systematizing this business as to render the working of it as nearly automatic or mechanical as possible, and thus eliminate as far as practicable the risk arising from the deficiencies of the human agency. In all systems worked by man this risk will be found. Better pay will procure better men, greater care and greater conscientiousness. Men laboring for a bare pittance and with little hope of advancement in the future do not usually cultivate these qualities to the highest point. Thus we are brought to one of the many points where the balance must be constantly sought between economy of expenditure and security of management. Each railroad officer must work it out for himself.

Operators should aim at a high standard of qualification and attention to duty. If the result is not greater remuneration in this service, their efforts may be rewarded by promotion in other directions. Reliable men are always wanted, and the consciousness of doing one's best is a source of satisfaction of more value than money. A careful study of their special work will develop a sense of its importance, leading to better attention to duties and preparation for advancement. Operators will therefore do well to make themselves masters of their business, rather than rest satisfied with a merely mechanical attention to prescribed methods, without an intelligent apprehension of their significance.

Telegraph offices should be carefully guarded against the intrusion of outsiders or employés off duty. Conversation or other interruptions may distract the attention at a critical moment and cause an operator to write an order incorrectly or allow a train to pass which he should stop.

CHAPTER IV.

There are some general considerations which it is important to bear in mind in the preparation and issuing of train orders. Some of these have been already pointed out. The circumstances under which they are to be acted upon render it of the utmost importance that there shall be nothing in their form or matter to obstruct in any way a clear and prompt comprehension of their intent. *No instructions should be included that are not strictly running orders.* Directions to take on or put off cars, or to change engines, or general instructions as to the management and stops of a train with reference to its traffic, are not properly included in such orders. Again, *the language in which the orders are expressed should be simple and unmistakable.* Simplicity implies brevity. Superfluous words or ambiguous terms or expressions should be carefully excluded. To avoid the use of anything of this character the precise form of expression should be determined on beforehand for all cases that can be anticipated, and strictly adhered to. This also renders the work of the Dispatchers uniform, and enables them to perform it with facility, especially if not greatly experienced ; and the trainmen become accustomed to the forms, and comprehend them at sight.

There are differences of opinion among practical men as to the propriety of including more than one transaction in the same order. Some reasons have been before urged against this practice. As men generally favor the practice to which they are accustomed, it is not easy to settle this question. A number of meeting points may be given in succession in one order more readily in the "single order" system than in the other ; and this is claimed as an advantage, and as better than giving the same on as many different pieces of paper. With an order, hastily and perhaps poorly and closely written on flimsy paper, to be read by a conductor in a storm or by the dim light of a hand-lamp, there is a good deal of risk that in a long order for several meeting-points something may escape notice ; a line may be skipped and a meeting-point missed. In the "duplicate" order the same danger would exist, and, in addition to the matter affecting the train receiving an order, it would get matter not at all affecting it. Thus, if A is ordered to meet B, and B to meet C, and both orders are included in one for the benefit of

B, the duplicate to A would include matter for C in which A has no concern, and that to C would have matter for A which he does not require. Circumstances might make it of some use for A to know where C is to meet B ; but burdening the order with this extraneous matter will be found usually to be a positive disadvantage and to cause much more work in transmission than giving each operation singly. The latter has been found to work entirely well in practice, and is theoretically the safer method. The conductor or engineman holding several of these orders arranges them in their proper succession, and each one as it is fulfilled is laid aside. It may be desired to change a meeting-place ordered, and, if this is included in an order containing several others, the change is not so readily made. The reasons would appear to be important for insisting *that each order should be ordinarily confined to a single transaction*, with slight exceptions, some of which are elsewhere adverted to.

The following is a sample of "duplicate" order actually and frequently given in practice on one of the principal divisions of the Pennsylvania Railroad. It is given to illustrate perhaps the least objectionable method of combining several movements in one order. It is compact, and is alleged to serve a good purpose. The principal objections to it are those above given.

PENNSYLVANIA RAILROAD COMPANY.
C. T. 202.

PHILADELPHIA DIVISION.

Telegraphic Train Order No. *14*

Superintendent's Office, West Philadelphia, *March 10th* 1888

To Conductor and Engineman

of 1st & 2nd No 6 Hy, at 1st & 2nd No 9 DY
1st & 2nd No 7 & 1st & 2nd No 3 Lancr.
1st No 6 and 1st & 2nd No 9 will meet at Branch Int.
1st No 6 and 1st No 7 will meet at Milladila,
1st No 6 and 2nd No 7 will meet at Conewago,
1st No 6 and 1st No 3 will meet at Elizabethtown,
1st No 6 and 2nd No 3 will meet at Kuhn,
2nd No 6 and 1st No 7 will meet at Branch Int.
2nd No 6 and 2nd No 7 will meet at Milladila,
2nd No 6 and 1st No 3 will meet at Conewago,
2nd No 6 and 2nd No 3 will meet at Elizabethtown.

31 C.

Painter	Hoffmaster	1st 9
Fulton	Conductor.	Miner	1st 7
Rattew		Kelly	2nd 7
Jacobs	"	Webber	1st 6
Ruch		mouth met	2nd 6
Baldwin	"	Suism	1st 3
Connell		manahan	2nd 9
Blankenbiler		Shultz	

Received at M. from *C.F. Dunlap* Opr., by *N.B. O'Krosky* Opr.

Made Conoka M. from *S.P.* Opr., by *Y.B.W.* Opr.

Conductor and Engineman must each have a copy of this order. See Rule 96.

An order *must not be taken to allow more than it expressly authorizes*. As, for instance, a train authorized by order to run in the time of another is not on this account to assume that it may run within the time of any other superior train which may be understood to have to keep out of the way of' the train whose right is curtailed. Each train must be governed in all respects by train rules with relation to every other train, excepting as distinctly provided in the special orders ; and as a necessary consequence of this, *no train should be permitted to run under the authority or protection of an order given to another.*

Every provision in an order should be held to be *in force indefinitely until fulfilled or annulled, or expired by some limitation in the order or in the rules.* In the orders delivered to those who are to execute them *no erasures, alterations, or interlineations should be permitted.* These tend to obscure the meaning and raise doubts as to accuracy. The writing should be clear and plain, the letters well formed and without flourishes. Orders must often be read in dim light or in storms, and when men are hurried, and they should not be required to decipher bad writing. Many orders have come under the author's notice which were defective in this respect. The following specimen is given, omitting

names that would indicate where it was issued.
The bad writing, the number of points covered
by the order, the difficulty arising from these,

and the flimsy character of the paper must
condemn the order as utterly unfit as a reli-
ance for the safety of life and property dependent upon its proper execution. The illus-

tration is not wholly satisfactory, for the reproduction of the order on smooth, white paper does not adequately represent the indistinctness arising from yellow paper, thin and crumpled, on which it was written, in common with so many train orders.

Orders should be identified by *consecutive numbers*, as is now usual. If the regular business requires a large number it is better to begin with No. 1 each day. As a precaution against the engineman overlooking orders, and as a means for properly taking care of them, a *clip should be provided for them on the engine, in a position to be readily seen by the engineman while attending to his duties.* This will avoid the necessity of his putting the orders where he may forget them ; and *with each on a separate paper* they may be arranged in proper succession and removed as executed, leaving always before the eye the next to be executed. The copies of orders retained by operators should remain in the book. These books and the copies that have been used by trainmen should be sent to headquarters for inspection. This will serve to indicate the manner in which the regulations are carried out, and the condition, as to legibility, etc., in which the orders are issued.

Forms of orders will be considered under "Forms."

2.

CHAPTER V.

Under the common practice there must be prepared at least three copies of each train order received for delivery. The conductor and engineman are each supplied with a copy, and the operator retains one. To make three several copies by pen and ink, as heretofore practiced by some, takes a good deal of time, and there is danger that they may not be all alike, and the time and risk are increased if more than three copies are required. To obviate this, the manifold system of writing has come into general use and with very great advantage. As used by many, however, it has serious defects. The tissue paper frequently used is very objectionable, especially the yellow variety. Messages written on it are quite difficult to read, especially in a poor light; it is easily crumpled, rendering it still more indistinct; it is difficult to handle in the wind, and it is easily damaged by wet. In the use of the manifold for some seventeen years the author found it entirely practicable to use an opaque white paper, of sufficient body to be free from the above objections and yet capable of giving

3

seven distinct copies with a good pencil of the hardness of No. 4 Faber. This is now recognized as the best and is prescribed in the specifications connected with the Time Convention rules.

Operators should not be permitted to receive orders on separate slips and copy them on the manifold, but should take the order down at once in the manifold-book. A sheet of tin placed in the book enables them to make all the copies perfectly distinct. Of course none but "sound" operators can do this. It takes but little more time and application to make a "sound" than a "paper" operator, and the advantage of the former is so great in this as well as in other respects in this service that it should always be required. Operators readily become able to take the requisite number of copies in manifold without the use of intermediate slips, and the risks of copying are thus avoided. When more copies are wanted than are made at the first writing they should be traced from one of the original copies. In the case of a general order, as in annulling a train, operators would usually make but one copy, and others required for delivery should be traced from this. Careful supervision should be had as to the actual practice of operators in the proper use of the manifold, and as to frequently changing the carbon paper to secure distinctness.

CHAPTER VI.

A careful record ought to be kept of each step in the issuing of an order, as well as of its exact terms. This record should be made on the original copies held by the Dispatcher, and by the operator who receives and delivers the order. The Dispatcher's copy should show who issued it, and both should indicate what operators were engaged in its transmission, and the time at which each step was taken, as well as the proper address, etc.

The Dispatcher's train sheet should constantly show the movements of the several trains, which should be promptly reported by the operators and recorded by them in the prescribed forms. A practical difficulty occurs in making the Dispatcher's record of all the steps in the issuing of an order, which it may be well to refer to here. When the Dispatcher is assisted by an operator, the most of the steps will be taken and recorded by the latter. They should be at once recorded on the original copy of the order, so as to leave nothing to be remembered or copied. Now, if the Dispatcher must write the order out in the book before transmission, the operator may have occasion to use the book at the same time for recording

steps then in progress with reference to other orders ; and if he does not, the passing of the books back and forth between them is inconvenient. It has, partly on this account, doubtless, become the custom with many for the Dispatcher himself to telegraph the orders without first writing them down, his operator taking them down as repeated back and writing them in the book of record. The operator thus has the book all the time in his hands. The objections to the Dispatcher transmitting orders himself are elsewhere considered, and it is designed here to point out a method by which the other plan can be pursued and the inconvenience referred to avoided. The Dispatcher is provided with a manifold-book and some loose sheets properly headed. With these, by the manifold process, he prepares two copies of the order, one in his book and the other on a loose sheet which he hands to the operator for use in transmitting. On this all the subsequent record is made by the operator, and at the close of each day all the orders for that day are fastened together and filed away. The numbers and manifold writing sufficiently identify the two copies if subsequent comparison is necessary, each being in fact an original. This method has the further advantage that the Dispatcher has by him all the time copies of orders he has issued, for reference if needed.

CHAPTER VII.

A method much used for signaling a train to stop for orders is to display a flag or light of suitable color, after receiving the direction to "hold the train." This is often done by holding the signal in the hand or placing it on the platform or ground or in some fixed place. If placed on the platform, without attendance, it is liable to be obscured or removed by persons about the place. If held in the hand of the agent or operator it is a poor arrangement for performing so important an office. The operator is usually required to report that the signal is displayed. He evidently cannot do this without leaving the signal unattended, and in fact when he is alone he must so leave it, as, after it is displayed, he must return to the office to receive the order, and he must also often be engaged in his office while expecting a train. It will frequently occur that trains will pass his station after he has received an order for some subsequent train ; in which case he must temporarily remove the signal, or stop a train which might otherwise not be required to stop. When this plan is used all trains that arrive before that for which the order is held are actually stopped. A serious accident occurred some years since from the hand-lamp going out as it was swung as a signal to stop a train for which orders had been received. The signal failed, and the train went on and collided with the opposing train. Lanterns and flags are the only available movable signals to be put in the hands of train and track men, but they should not be relied upon where anything better can be used. The evils attending this use of hand signals are so manifest that the practice is fast disappearing, and the reference to it here may before long be only a reminder of what has been done.

A signal for this purpose should be distinctive and of the most substantial character. A fixed signal manipulated from within the office is greatly to be preferred. Several such have been devised. The signal should be such as to be distinctly seen at proper distances; it should be as little as possible liable to confusion with other objects, and it should be an adornment rather than a disfigurement to the landscape in which it forms a prominent feature. The most satisfactory signal within the author's knowledge is the simple semaphore arm, extending horizontally from a post and showing a red light to signify "stop," and inclined and showing a white light to signify the opposite, and operated by a handle within the telegraph office.

Much discussion has been had in the past as to whether a danger signal, which this pre-eminently is, should stand normally at safety or danger. The earlier practice favored the former, as indicated above, the absence of a signal, in the plan described, being the rule. In more recent years the weight of opinion has been that in all systems of danger signals the normal position, and that to which such signals should automatically move, is that indicating danger. So arranged, the indicator will always be in a position to stop trains unless it is moved to another position to show that there are no orders for them. It becomes a stand-ing order to "hold," and, when an order is forwarded for a train, the fact of its receipt requires that the signal be simply left in its normal position and the train thus stopped. It will be then the rule and the habit of trainmen to observe all these signals and to stop when they are not placed, on their approach, in the position permitting them to proceed.

The rules of many railroads still indicate a usage contrary to this. The lamp, flag, or other stop signal is displayed only when a train is to be stopped for orders. It appears that under some circumstances, especially where the duties of the agent

and operator are performed by the same
person, the telegraphic duties being comparatively small, it is thought better to retain this
method, and the rules of the Time Convention
were so framed as to provide for either, leaving the choice to those concerned. Under the
"normal at danger" plan, when an order is
received in advance of the arrival of the train
for which it is designed, and has been properly
verified and prepared for delivery, it remains
in the hands of the operator until the train arrives, the signal showing "stop." If, in the
mean time, other trains pass for which there
are no orders, the signal must be placed, as
they approach, so as to indicate that they may
pass. But there is then the danger that the
operator may inadvertently allow the train to
pass for which he has an order. This has actually occurred, and should be provided against.
This should be done by requiring that, as soon
as an order for a train not arrived is ready for
the signatures of the trainmen, or for delivery
when signatures are not taken, the copies designed for them shall be removed from the
book, folded, and marked with the train number, and put in a designated place and in such
position that the signal handle cannot be
moved without the eye and hand being directed
to the orders. This is readily effected by a
rack to hold the orders placed on a small door

closing by a spring and catch over the handle by which the operator moves the signal. The handle cannot be moved without unfastening the door and so opening it as to bring the orders, which are on it, under the eye and hand of the operator. This precaution may appear trivial, but while it is of great importance to adopt such routine that its mechanical performance will lead to a correct result, it is equally important to interpose such obstacles as are necessary to prevent a mechanical inadvertence that may lead to disaster. The same kind of risk exists in the use of block signals, and several plans have been used to obviate it by suitable mechanical means. In the other use of the train-order signal there is, to a certain extent, the same liability to this unconscious movement when it has been placed at danger, and a like precaution is needed to guard against it. It often happens that there are orders on hand for several trains. A definite place for them prevents their getting mixed with each other or with other papers ; and removing them from the book avoids the necessity of leafing them over to find the particular order which men may be waiting to sign, and possible mistake in getting the wrong order.

The only reason of apparent moment that could be assigned for leaving the orders in the book is that the trainmen may sign all the copies. There does not appear to be any good reason for requiring their signatures on their own copies, and the manifold writing by them would be unsatisfactory. Again, it will often happen that more than one train is to receive a copy of the order, in which case the same signatures are not wanted on all the copies. The point here urged as of paramount importance *is that the order itself shall be interposed between the operator and the instrument by which he might give a signal permitting a train to pass improperly.* In this view the discussion of the point is pertinent to the subject of "Signal." It may be added that the final indorsement of "complete" after signature on each copy takes but a moment, and perhaps no longer than a careful writing of it over several copies in the manifold-book; and as the men should read and compare their copies before the final steps, it is difficult to see how they could do this properly if the orders remain in the book.

The train-order signal should be used for no other than its legitimate purpose. It will not be inconsistent with this to use it for holding a train the required time after the passage of another train in the same direction.

Upon some roads, trains passing while the stop signal is shown receive a "clearance" card stating that orders in hand are not for them. This is included as a part of the plan presented in the Time Convention rules for the use of the signal with its normal position at safety. It would seem to be necessary with this method; and in any case where it can be used it is a valuable precaution, the only objection being that it requires the stopping of fast or heavy trains which it might be quite objectionable to stop. This would seem, however, to be proper for any train stopped by the signal for time.

Where the plan is adopted of keeping the train-order signal normally at safety it should still, as in the other system, be so arranged that it will move automatically to danger if any of the mechanical parts fail. If this is not done and dependence is placed on fastening it at danger, the fastenings or some of the connections may fail and the signal move to safety without the fact being observed. One important advantage of the other plan of using the signal is that it is never at safety excepting when held in that position by the operator. Where the usual position is safety it cannot be arranged for the operator to actually hold the signal while it occupies the danger position.

CHAPTER VIII.

The transmission of orders will be taken to include all the steps after preparation by the Dispatcher until final delivery.

These are :

1. Telegraphing the order to the stations to which it is to be sent.

2. Writing down as received.

3. Repeating it back to the Dispatcher.

4. The response of the Dispatcher indicating that it is correctly repeated.

5. The acknowledgment of this response.

6. Comparing copies of the order with the persons to whom it is addressed, and taking their signatures.

7. Telegraphing the signatures to the Dispatcher's office.

8. The Dispatcher's reply, acknowledging the receipt of the signatures, and indicating that the order may now be delivered.

9. The indorsement of this reply on the order.

10. The delivery to the trainmen.

Some Dispatchers prefer to personally telegraph their orders, having an assistant operator to copy them as transmitted or as repeated, and to perform the subsequent work of verification, record, etc. Those who are accustomed to transmit their own orders strongly contend for that practice. Those who pursue a different course are equally strong for theirs. In arranging for those, at least, who have not become wedded to any particular method, general consideration should govern. If contests or inquiries arise on the wire when the Dispatcher is sending, time is occupied which he may very much need, and where the amount of work is large it will leave the Dispatcher more at liberty to attend to his special duty if he simply prepares his orders and hands them to an operator for the subsequent steps, and this is by some carefully insisted upon.

The Dispatcher's duty is not simply to direct each movement as the exigency arrives. He should be constantly on the alert to provide as far as possible in advance for the arrangments necessary for keeping his trains moving, and his mind should be free from anything that may interfere with this. Attention by him to the merely mechanical duties detracts from his usefulness and the benefits which the road should derive from the talents which are supposed to fit him for his position. Some points connected with this subject are referred to in Chapter VI. Whether sent personally by the Dispatcher or by an operator from a written sheet, the order should, whenever practicable, *be transmitted simultaneously to all the offices to which it is to be sent.* Ordinarily this will be to but two offices. An order annulling a train may have to be sent to all the offices on the division. The simultaneous transmission is a most valuable safeguard and a saving in telegraphing only practicable with the duplicate order. It has been urged as an objection to the duplicate order that where agents act as operators their duties as agents may sometimes interfere with their attendance as operators when wanted for simultaneous transmission. This furnishes no ground for objecting to this form of order, as simultaneous transmission is not essential, and it is only necessary in such case that the precaution be observed of sending first to the train of superior right.

On calling an office a special signal should be used to indicate that a train-order is to be sent. The numerals 31 or 19 are now generally used for this purpose, the former for orders to be signed by the trainmen before delivery and the latter for orders to be delivered without such signature. After this signal the word "copy" should follow, with a number indicating how many copies are to be made. This may be omitted when three is the number required, that being the most usual. If the system in use does not provide that the train-order signal shall stand normally in the "danger" position, the operator who is to receive the order must, at this point, place it in that position and report that he has done so. He then prepares his manifold-book for the requisite number of copies and takes the order down as sent, with the proper address for his station, immediately repeating it back word for word, *reading from the order as actually written on the paper to be delivered*, and not from a slip to be afterward copied. A "paper" operator should write the order in manifold before repeating. Some defer the repeating until the signatures of the trainmen are to be reported. But it is on many accounts preferable to repeat and verify the order at once and before signatures are taken, even if the trainmen are present. It assures its accuracy before they have read and signed it. The repeating operators can listen to each other better than if they repeat at different times, and the sender of the order can better attend to its verification while the original lies before him. There will also be less detention to trains if the repeating is done before their arrival. The importance of this will further appear from the consideration elsewhere of the effect of an order where the telegraph fails after but one train has received and proceeded on it.

The relative succession in which the offices are to repeat should be fixed by rule or usage, to avoid doubt or conflict. It is better that the repeating be done in the same succession as that in which the several offices are addressed. This assures the repeating first by the office receiving for the superior train. As a valuable precaution against error, *each should be required to listen while the others repeat.* An operator has been known to hear the name of a meeting-place correctly, write it down incorrectly in the order and repeat it back correctly. If he had looked at his copy as the other repeated, he would probably have noticed his error.

In this connection it may be observed that too much importance cannot be attached to the cultivation of a careful habit in telegraphing orders. A certain degree of rapidity in handling the key is not inconsistent with distinctness, but the latter should never be sacrificed to haste and a hurried and careless style of telegraphing should never be permitted.

The operator in the Dispatcher's office should carefully observe each word as repeated by each, to make sure that all is repeated correctly. Some observe the commendable practice of underscoring each word as repeated, thus making sure that their attention is not withdrawn. If the Dispatcher transmits his

4

orders himself and his copy for record is made as the order is repeated, as is the practice of some, his copy can hardly be said to be an original. It may vary from what was sent or designed to be sent, and his operator taking it down has not the opportunity of checking as above, and may himself make a mistake in receiving it. All offices required at the time to repeat an order should do so before the Dispatcher replies. The signal for this reply now generally used, and adopted for the "Standard" Code, is "O K." This is given simultaneously to all, naming each, and each should acknowledge it. It is important that the Dispatcher should know that each has received the "O K." It is not necessary that the Dispatcher personally authorize this reply. It may be properly done by his operator who has watched the repeating. Where the order is not repeated back until the signatures are obtained and sent with it, the response, "O K" and sometimes "complete" is used to cover the whole, but where the practice herein recommended is pursued, the use of two signals is necessary, "O K" being the first. The time at which the order is sent and "O K" given should be noted on all the copies, with the initials or signals of the operators sending and receiving, and the name or initials of the superintendent. The order is then ready

for signature and delivery, and, if the train
for which it is designed has not arrived, the
train copies should be removed from the book,
folded and marked on the outside with the
train number, and placed in the rack provided,
as indicated under The Train-Order Signal.

Practice has varied very much in the method
of delivering orders. Some have simply had
them authenticated by repeating back as above,
with perhaps the proviso that the trainmen compare their copies with that of the operator, and
in some cases sign for them. The transmitting
of signatures has not in all cases been required.
Many rules, especially those of early date,
appear to be based on the idea that the whole
process of sending, verifying, and acknowledging an order is to be continuous and while the
train is at the station. Much that appears in
some rules gives the impression that either
this idea prevailed or that the phraseology
used in connection with it was retained while
the practice had changed. On a busy road it
would certainly be impracticable to carry out
this idea, and it is not now usually attempted.

In early days of train telegraphy, when orders were not prepared with the precision of the
present day, it was the custom to add to the
order the phrase "how do you understand?"
This came to be represented by a signal, the
most generally used perhaps being the numeral

"31." The reply to this, preceded by "we understand we are to," represented by "13" or other numeral, was required to be written out by the trainmen as their "understanding." This was probably in most cases a verbatim copy of the order. Whether this was actually done by the conductor and engineman is doubtful. Some allowed the operator to do it. With the definite forms of orders now used and well understood, there is certainly no necessity for men to write out their "understanding." The manifold copies, authenticated by repeating back and compared by reading aloud, which also serves to impress the order on the men, must certainly be better than anything written by or for them. There would seem to be no reason for perpetuating a fiction by referring to the repeating of the order as the "understanding" or by the use of "31" and "13" in their original sense, when the question and answer which they represent are no longer designed to be used, and this practice and the expressions which arose under it have almost entirely given place to the improved methods.

Following, then, the practice here recommended and now generally used, the message has been placed in the hands of the operator and its verbal accuracy assured, and the train-order signal being in position to stop the train, the conductor and engineman understand that on

arrival they are to go to the office " for orders."
One of them (or the operator) should read the
order aloud while each looks at his copy, the
object being *to guard against a hurried reading of the order, to acquaint them fully with
its exact terms, and to impress its purport
upon them.* It is to be hoped that no man
would willfully disregard a train order, but
there are many who would proceed upon a
hasty examination or none at all, if permitted
to do so, and perhaps on a wrong impression
as to what it directs to be done.

The order having been thus read and compared, the signatures should be taken on the
operator's copy. From the many rules forbidding operators to sign for trainmen, and conductors for enginemen, it would seem probable that
this is sometimes done. This is a practice
which no considerations of convenience can jus-
tify. Personal signatures should be insisted
upon. Without this there is danger that men
will hastily "grab" an order and fail to get
its meaning. Time is well spent in securing
their particular attention to it, and their signatures attest that this has been done.

There is much difference of opinion as to
whether it is important to take the signature
of the engineman. Much time is often lost by
taking him from his engine, particularly on
very long trains, and some think that the pur-

pose is as well served by having his copy delivered to him by the conductor. In the latter plan there is some danger that the attention of the engineman may not be particularly called to the purport of the order, and for this reason the author believes that the practice is best where both signatures are required. The Time Convention code leaves the choice optional.

The signatures having been obtained, the Dispatcher is to be advised, by their transmission to him, in connection with the number of the order signed for and the train number or designation. The reply that all is satisfactory, authorized by the Dispatcher personally, is then to be given in some prescribed form. The word "complete" has been adopted in the "Standard Code," superseding "correct," which was formerly used.

The selected word should be written on each copy, with the exact time at which it was given. The order may then be delivered, and the train order signal so placed as to allow the train to proceed. If the Dispatcher's office is also used as an office for delivering orders, the same formalities in delivery should be observed as at way offices.

It will sometimes occur that an order must be sent to a disabled or other train away from a telegraph station. It must, in that case, pass through additional hands, and great care is necessary to guard against error. The conductor or messenger who carries the order should be made accountable for its delivery in proper form, by himself signing for it and getting "complete." The order being addressed to the conductor and engineman of the train "in care of" the messenger selected, the latter should be furnished with an additional copy, on which he is to take the signatures of the conductor and engineman, as if they were at a telegraph office. This copy should be delivered as soon as practicable to an operator, who should forward the signatures, completing the process.

Although when these paragraphs were first written the method of transmission described did not correspond entirely with any practice that might be termed general, it agreed in essential points with the practice upon several roads where most careful consideration has been given to the various risks in train dispatching and to methods for avoiding them. The process detailed indicates the points to be guarded, and furnishes what has proved a practicable and satisfactory method, and corresponds with the regulations now being rapidly adopted on our principal roads.

The rules should determine the course to be pursued if the telegraph fails during the process of transmitting an order. If this occur

before its correct reception is assured by repeating back and giving and acknowledging
"O K" for any office concerned, the process
is not sufficiently complete for the men
of a train at such office to be allowed to
sign for and act upon it. If, therefore, communication is not quickly restored it is perfectly safe and proper to provide that an operator shall permit a train, in such case, to proceed '
on its schedule rights without orders. If, on the
other hand, "O K" has been given and acknowledged, the correct reception of the order
is assured, and a period is reached when the men
of a train may, and often must, be permitted,
on arrival, to sign for and act on the order
before the arrival of the other at the point
where the order is awaiting it. If the men of
one train have thus proceeded, and the other
on arrival cannot be communicated with,
it would be obviously unsafe for it to proceed upon the order awaiting it for which
signatures cannot be transmitted, because,
although the opposing train may be on the
way to execute the order, this is not known
to the train that is cut off from communication. It would therefore be improper for it
to proceed either in accordance with the order
or on schedule rights. It would appear, therefore, that an order wholly or partly sent by
the process detailed, and for which "O K" can-

not be given and acknowledged by reason of the telegraph failing, should not operate to hold the train addressed, but that an order for which " O K " has been given and acknowledged should have this effect. The rule should therefore be *that, after "O K" is given to an order and acknowledged, the train to which the order is addressed shall not be permitted to pass until the signatures are transmitted and "complete" obtained*, or until the train can be communicated with by the Dispatcher. This is based, of course, upon the presumption that the plan is followed of assuring the accurate transmission for both trains, and that each operator has acknowledged the " O K " before "complete" is given to either. The delays arising from the operation of this rule cannot be frequent, and it is better to submit to these than to run the risk involved in a different course.

In the use of the "19" order, to which the signatures of the trainmen are not taken, the order becomes of effect only when "complete" has been given and acknowledged ; and until this is accomplished it should be treated as of the same effect as a "31" order for which " O K " has not been given and acknowledged.

If the practice is followed of delaying the repeating of the order until the signatures are obtained and sent, then the presence of the order in the operator's hands should serve to hold either train if the telegraph fails, as neither can know but that the other train has received the order and proceeded on it. It must be seen, however, that there is some risk in depending on a train being held by the mere presence of an order, the correct reception of which has not been fully acknowledged, as the receiving operator may even have made an error in receiving the number of the train for which the order is designed; and this offers an additional reason for repeating back at once on the receipt of the order. These considerations as to the holding effect of an order when the telegraph fails, do not, of course, apply to a general order, as one annulling a train, until such order is specially addressed to a train. It should be understood that operators hold trains a reasonable time for the resumption of communication broken during the transmission of orders.

It is important that the holding effect of an order not signed for should be clearly understood, so that the Dispatcher may run trains with confidence against a train so held.

A careful Dispatcher will observe that the inconveniences arising from a train being held by the incomplete transmission of an order will be greater as the distance is greater between the point to which the order is sent for delivery and the point where it is to take effect.

CHAPTER IX.

Many books of Rules have borne evidence that the ability to construct rules is not always commensurate with the many other gifts of successful railroad officers. To know what is to be done and how is one thing, but it is quite another to express the intention clearly and concisely. A scholar might present the subject in precise and grammatical form, and yet fail to so render it as to make it plain to practical men of limited education ; and yet, while the language must be clear to the untrained mind, there should be no expressions that are not within the bounds of rhetorical propriety. The evident difficulties surrounding the subject render more conspicuous the admirable results of the work of the able committee of the General Time Conven tion in the production of the "Standard" code of train and telegraph rules contributed by that body to the railroad service. To have produced a set of rules that should be accepted for general adoption, in which so few deficiencies have been pointed out, is a work worthy of the highest commendation. Under the operation of these rules will disappear the uncertainty often appearing in anxious inquiries by "Conductor" or "Train-Master," in the railroad papers, as to how this rule or that order is to be understood under given circumstances. There will be fewer occasions for trainmen to reconcile conflicting regulations and fewer cases of "doubt," in which to "take the safe course and run no risks."

No one, however, feels that entire perfection has been reached, in practice or statement, or that even in the near future, additions or changes may not be found desirable ; and, as methods of operation improve, scope will doubtless still be found for fresh talent in the production of regulations for new combinations of circumstances as well as improvement in those prepared by earlier hands. ˉ

The Telegraph Rules of the Time Convention, adopted October 12th, 1887, are here given, with some discussion relating to them. In considering these rules mention will necessarily be made of points referred to on previous pages and which are here embodied in form for practical use. This necessarily involves some apparent repetition. The rules are here designated by the numbers given to them by the Time Convention Committee; and it may be here stated that, in conformity with the method followed in the Time Convention train rules, the term "time-table" is herein applied to the issue governing the movements of all regular trains, while "schedule" is used to designate that part of the time-table which applies to any one train.

RULE 500.—Special orders directing movements varying from or additional to the time-table will be issued by the authority and over the signature of the Superintendent. They are not to be used for movements that can be provided for by rule or time-table. They must not contain information or instructions not essentially a part of them.

They must be brief and clear, and the prescribed forms must be used when applicable ; and there must be no erasures, alterations, or interlineations.

This rule indicates the proper function of a Telegraphic Train Order, the authority under which it is to be given, and the essential features of its construction, with the requirement that the prescribed forms are to be used when applicable. While in the fixed forms provision is made for the majority at least of the cases likely to occur, occasions will doubtless arise when other forms or modifications of these will be required. It is therefore important that the principles on which these forms are to be constructed be distinctly stated. The provisions as to how orders shall be issued and as to the use of the forms, when applicable, and the absence of alterations, are all necessary as tending to secure uniformity and accuracy. The following note, attached by the Time Conven-

tion Committee, emphasizes a point hereinbe-
fore dwelt upon as of great importance :

[Note.—On Roads whose organization provides that any other
officer than the Superintendent shall direct train movements,
the official title of such officer may be substituted in the above
rule. The Committee considers it essential, however, that but
one person's signature should be used in directing train movements on any dispatching division.]

Rule 501.—Each order must be given in the same words to
all persons or trains directly affected by it, so that each shall
have a duplicate of what is given to the others. Preferably
an order should include but one specified movement.

Here is determined the feature essential to
the "duplicate" system, viz., that the order
shall be "in the same words" to all concerned;
and the preference is here given to the point
urged by the author, of covering but one movement by an order.

Rule 502.—Orders will be numbered consecutively for each
day as issued, beginning with No. 1 at midnight.

The use of numbers for orders serves to
identify each order and to indicate the priority
of issue.

Rule 503.—Orders must be addressed to those who are to
execute them, naming the place at which each is to receive
his copy. Those for a train must be addressed to the conductor and engineman, and also to a person acting as pilot.
A copy for each person addressed must be supplied by the
operator.

The requirement here that orders shall be
addressed to those who are to execute them
might seem superfluous but for some former

looseness in this respect and the necessity for exactness in prescribing each step in the process of issue. The address, including the place of delivery, is necessary as indicating, in simultaneous transmission, which operators are to receive for those respectively to whom the orders are sent. The introduction of the Pilot here is valuable. As the one under whose special direction the train is for the time being, he should be directly informed of orders controlling its movements. The conductor and engineman who are in charge of the train subject to his control, are also necessarily advised. The relations of the Pilot to the train are much the same as those of the pilot to a vessel of which he has control for the time being. He is placed there because of his having special knowledge, not possessed by the conductor and engineman, of circumstances which necessarily affect the movement, and has entire control of the train in this respect. He may or may not be an engineman. He may or may not run the engine. He, however, is to say when it may or may not run, and is the person by whose authority the movements are to be regulated with reference to the signals and the physical features of the road and with respect to other trains as well as the established rules. He does not assume the duties of the conductor as to those things which are purely

local to the train, and the brakemen and fire-
man are properly held to be under his orders
through the conductor and engineman. The
trainmen are not, by the presence of the Pilot,
relieved from the usual obligation to protect the
train and perform other duties connected with
it or required by the rules.

Rule 504.—Each order must be written in full in a book
provided for the purpose at the Superintendent's office; and
with it must be recorded the names of trainmen and
others who have signed for the order, the time and signals,
showing when and from what offices the order and responses
were transmitted, and the Train Dispatcher's initials. These
records must be made at once on the original copy, and not
afterward from memory or memoranda.

The requirement here as to the record of
each order in a book is usually now fulfilled
by the preservation of a manifold copy in the
book in which the blanks are bound. . This, in
fact, is the method contemplated, although the
rule is so drawn as to admit of other methods.
The record of the various points specified is
requisite for a complete history of each transaction.

Rule 505.—The terms "superior right" and "inferior right"
in these rules refer to the rights of trains under the Time-
table and Train Rules, and not to rights under Special
Orders.

This rule is rather an authoritative statement of a logical conclusion from the facts,
but very properly gives this prominence to a
point that must be constantly borne in mind.

When the rights of trains are reversed by an order, as is usually the case, the inferior becomes for a time the superior, and this definition emphasizes this. In this connection it may be again noted that a very important and necessary part of the training of those engaged in operating the railroad telegraph is the acquisition of an intimate knowledge of the rules governing the rights and movements of trains when acting independently of telegraphic control. The legitimate use of the telegraph is to facilitate movement when, under the unaided operation of the rules, there might be delay, and to give preference, for special reasons, to trains which, under the rules are inferior. An exact knowledge of the effect of the rules, and what may be done by trains under their provisions, is therefore important, so that there shall be no unnecessary use of special orders, and that those used shall be the most appropriate to the circumstances.

RULE 506.—When an order is to be transmitted, the signal " 31 " (as provided in Rule 509) or the signal " 19" (as provided in Rule 511), meaning "Train Order," will be given to each office addressed, followed by the word "copy," and a figure indicating the number of copies to be made, if more or less than three—thus, "31 copy 5," or "19 copy 5."

This rule begins upon the details of transmission and is the first in which mention is made of the special signals " 31 " and " 19," signifying "train order," the use of which is

5

more fully indicated later on. We have here the first step in the methodical plan of transmission prescribed in these rules, preparing the operator for the reception of the order and informing him of the number of copies for which he must prepare his manifold sheets. As three is the number most usually required, the omission of this number economizes telegraphing. In the same case the word "copy" might as well be omitted.

RULE 507.—An order to be sent to two or more offices must be transmitted simultaneously to as many as practicable. The several addresses must be in the order of superiority of rights of trains, and each office will take only its proper address. When not sent simultaneously to all, the order must be sent first for the train having the superior right of track.

[NOTE.—On roads which desire the operator at a meeting-point to have copies of the order, the several addresses will be, first, the operator at whose station the trains are to meet and next in the order of superiority of the rights of trains.]

This rule brings us to the transmission of the order and requires that it be simultaneous as far as possible. This is a safeguard possible only with the duplicate system. Here also the priority of transmission to the superior train is insisted upon. In addition to other advantages, the systematic naming of the superior train first calls the attention of operators to the relative superiority of trains. The principle involved here is elsewhere recognized. The note attached by the Time Conven-

tion Committee has reference. to the arrangement which some prefer of sending a copy of
the order to the operator at the meeting-point
in addition to the copies sent to other points
for delivery to the trains.

RULE 508.—Operators receiving orders must write them out
in manifold during transmission, and make the requisite
number of copies at one writing or trace others from one of
the copies first made.

This rule directs the use of the manifold
writing and practically dispenses with any
record book other than that in which the manifold copies are preserved.

This is one of the most important improve-
ments over the old methods. In the early days
of telegraphing and with some to a comparatively recent period, each copy of an order was
written separately, occupying much time and
involving great liability to error in transcrib-
ing. Now the perfection of the manifold admits of making at one writing all the copies
usually required. If additional copies are
wanted, their exactness is assured by tracing
from one of those made at the first writing It
must be observed here that the rule does not
permit an operator to take the message down
on a separate sheet and make his manifold
copies afterward.

RULE 509.—When an order has been transmitted, preceded
by the signal "31," operators receiving it must (unless otherwise directed) repeat it back at once from the manifold copy,

and in the succession in which their several offices have been addressed. Each operator repeating must observe whether the others repeat correctly. After the order has been repeated correctly by the operators required at the time to repeat it, the response " O K," authorized by the Train Dispatcher, will be sent simultaneously to as many as practicable, naming each office. Each operator must write this on the order with the time, and then reply "i i O K," with his office signal.

Those to whom the order is addressed, except enginemen, must then sign their names to the copy of the order to be retained by the operator, and he will send their signatures to the Superintendent. The response "complete," with the Superintendent's initials, will then be given, when authorized by the Train Dispatcher. Each operator receiving this response will then write on each copy the word"complete,"the time, and his last name in full ; and will then deliver a copy to each person included in the address, except enginemen, and each must read his copy aloud to the operator. The copy for each engineman must be delivered to him personally by ———, and the engineman must read it aloud and understand it before acting upon it.

[NOTE.—The blank in the above rule may be filled for each road to suit its own requirements. On roads where the signature of the engineman is desired, the words "except enginemen " and the last sentence in the second paragraph may be omitted. See also note under Rule No. 500.]

[Individual operator's signals may be used when desired in addition to office signals, as here and elsewhere provided for.]

In this rule are given in detail the steps to be taken after the order has been transmitted, this rule having special reference to the orders for which signatures of trainmen are to be taken, known technically as the " 31 " order. Much of the efficiency of the telegraph, as well as the safety of operation, depends upon the careful drill of operators in this respect and strict adherence to the requirements of the rule. Repeating back at the time of receiving may be properly omitted under the direction of the Dispatcher, in case of a general order, as one annulling a train. This would be sent to all stations but not necessarily delivered at all, and therefore repeating back at once from all would unnecessarily occupy the wire. Other cases may arise where the repeating may be postponed. In repeating, however, the requirement that it be done from the manifold copy should be carefully complied with. Reading, word for word, from the copy actually to be delivered is one of the most important precautions against mistake. The succession in which offices are to repeat is prescribed, so that all shall understand it, and it is so fixed that the repeating shall be done in the order of superiority of trains addressed. As a repeated order for which the "O K" has been given and acknowledged serves to hold the train addressed, this secures the superior train at once.

The requirement that operators observe the repeating by each other is a further valuable safeguard.

The next step, that of transmitting the "O K," is now prescribed in the same methodical way and its acknowledgment provided for. Without this acknowledgment the Dispatcher could not be sure of the train being held, and it is quite important, although not directed

in the rule, that the acknowledgment of the
"O K" should be made by the different offices
in the succession in which they were addressed.
This brings us to the point where the order is
fully in the hands of the operator and becomes
operative to a certain extent, as is seen in Rule
510. The train for which an order has thus
been sent may not have yet arrived. By the
rule, however, the signal is displayed to stop
the train, and when it arrives the conductor
(and the engineman if required) must go to the
office and sign for the order. The signature
(or signatures) must then be telegraphed to the
Dispatcher's office, and when found correct the
final response, "complete," is given, signifying
that all the steps in telegraphing have been
taken that are necessary before delivery. It
still remains for the receiving operator to record the "complete" on the order, with the
time and his name, all of which are important
for the completion of a paper which involves
the safety of human life. It is still, however,
possible that those who are to use this important paper may fail to observe its full signification, and it is therefore provided, as a final
precaution, that each one who receives it shall
read it aloud to the operator, who has his own
copy before him. This is better than reading
by the operator to the trainmen, as they might
not listen attentively, while they can hardly

fail to note the signification of words which they themselves read aloud.

The notes appended by the Time Convention Committee point out modifications which may be made with respect to certain points in which difference of practice prevails and which do not affect the essential features of the plan.

The author believes that the weight of sentiment is decidedly in favor of taking the signature of the engineman as well as that of the conductor for the order, unless controlling circumstances prevent.

RULE 510.—For an order preceded by the signal "31," " complete " must not be given to the order for delivery to a train of inferior right until " O K " has been given to and acknowledged by the operator who receives the order for the train of superior right. Whenever practicable, the signature of the conductor of the train of superior right must be taken to the order and " complete " given before the train of inferior right is allowed to act on it.

After " O K " has been given and acknowledged, and *before* " complete " has been given, the order must be treated as a holding order for the train addressed, but must not be otherwise acted on until "complete " has been given.

If the line fails *before an office has received and acknowledged* " O K" to an order preceded by the signal " 31," the order at that office is of no effect, and must be there treated as if it had not been sent.

[NOTE.—On roads where the signature of the engineman and pilot is desired, the words " engineman and pilot " may be added after the word " conductor " in the first paragraph of Rule 510.]

Rule 510 presents a requirement of very great importance in prescribing that " complete " shall not be given for the inferior train until " O K " has been given and acknowledged for the superior. The reason for this is apparent from the following considerations : When "complete" has been given, the train receiving an order on which it is indorsed may at once proceed to the execution of the order. If it has rights given to it against a superior train, it is of the highest importance that the latter shall be informed of this before it can proceed to a point where the order may bring the inferior into conflict with the rights of the other. After "O K" has been given and acknowledged for the order at the point where the superior train is to receive it, the order "holds" the superior train, as provided in the second paragraph, and it is only then safe to permit the inferior train to proceed, by giving for it the final word "complete." It would be still better if in all cases the signatures of the men of the superior train could be taken before the other is permitted to act on the order. The rule requires this "whenever practicable." It is, however, often not practicable on account of the varying and often considerable distances between telegraph stations, the varying speed of trains, and unforeseen and unpreventable delays. It is doubtful whether any reasonable expenditure in increasing the number of offices would admit of absolute compliance with such a requirement, but it is quite true that any expenditure at all ap-

proaching what this would require would
be much beyond the ability of the majority of railroads. It is also true that, at least
without enormous additions to the facilities,
a strict requirement of this kind would interfere with the movement of trains to an extent
that the patrons of the roads would never
agree to. If the plan provided in the rules
really involves any risk in this respect, it is
one which cannot be avoided in the present
state of financial ability and of the means of
moving trains.

The closing paragraph of the rule provides
for the contingency of the failure of telegraphic
communication at a critical moment in the
transmission.

An order may have been fully received by
an operator, but, if the telegraph fails before
he can repeat it back and be informed by the
Dispatcher that it is "O K," it would not be
safe to use it. Neither is it proper that it
should have any effect whatever until the Dispatcher is assured, by the acknowledgment of
the "O K," that it has been received. When
an order has been transmitted and is altogether
in the hands of the operator, there is the
chance that he may have written down some
important word incorrectly. Hence the requirement that he repeat it back. This, if carefully performed, assures the Dispatcher of the

verbal accuracy of the message as the operator
has it, and the Dispatcher admits this by the
response "O K." He must now act, with reference to this train, as if it were held at the
point at which it is addressed. But he cannot
assume this until he is assured that "O K"
has been received. This is by the required acknowledgment.

If communication absolutely fails before the
completion of this process, all that he has done
goes for nothing unless communication is
quickly restored. It is of the utmost importance that the Dispatcher know what will
or will not be done by a train to which an
order has been addressed, as this knowledge
guides him in giving other orders. It would
not be proper, even, to assume that a
train would be held by the presence of an
order addressed to it unless the accuracy of
the order is assured, for an error may have
occurred in receiving the address and the
wrong train number may have been noted.
Nor will it do for a train to proceed regardless
of an order addressed to it when the whole
process of transmission cannot be completed,
unless the rule authorizing it is made to
specify the precise point in the process of
transmission when this may be permitted.
It is also of equal importance that, in the
absence of telegraphic communication with

a train, the Dispatcher can depend upon the fact that it will act in accordance with the rules, notwithstanding a partial transmission of an order intended to control its movements. Briefly, he must know whether the train retains the right to proceed or not, and under what conditions, or he cannot intelligently direct other trains with reference to it. The question how long a train should wait for communication to be restored must depend upon so many circumstances that no rule can be given. The "break" may be but momentary or it may last for hours. The train may have just time to get to a regular meeting-place, at which, if reached in time, it may have to lie for belated trains. Rules must fail here to indicate what is best to be done, and often the best judgment is no guide. Whatever is determined on may involve delay. It should never involve danger.

There is a plan in use on several prominent roads by which it is claimed that the objectionable feature in Rule 510, represented by the phrase "whenever practicable," may be eliminated. Under this plan there is added an "advance" order, issued to the superior train, directing it to stop "for orders" at a point where it is intended to deposit for it the duplicate of a meeting or other order on which an inferior train is to be permitted to proceed from some other point before the order is received by the superior train. By this plan the superior train is "held" before the inferior is allowed to act on the order, and thus far the risk is avoided of the superior being improperly allowed to pass the point where the duplicate order is to be placed for it. It is claimed that a considerable experience has demonstrated that this plan is feasible and secures the object in view, and that with it the rule of always first securing the superior train may be made absolute. Experience is one of the best of teachers, and few theories can be taken as proved without it, but even imperfect methods may produce good results under careful management, so that experience alone is not sufficient for determining the merits of a system.

The purpose of the plan in question, to "hold" the superior train before giving orders against it is good, and what all wish to accomplish. This idea gave rise to the "hold" order of the older methods of train dispatching and it has been suggested that under the advance-order plan there is danger of a relapse from strict adherence to the duplicate method. Careful supervision may prevent this.

If the advance order is invariably given, operators may get to depending on it rather than on their own care for stopping trains at points where duplicates are deposited. This is a point to be carefully considered and on which the railroad fraternity will be by no means agreed. Two things are depended on. If one fails we have the other. Many hold that this is better than to rely on one alone. Many, again, maintain that, where the responsibility is thus divided, each party may depend on the other and both fail, while, if there is but one, his sense of responsibility is quickened and the result is better. In view of the difference of opinion on this point it may be said that if this be the only point in the consideration of the advance order it may be given a trial.

If it is to be tried, then we must see that there are no exceptions to its use. The Dispatcher must always anticipate possible contingencies long enough ahead to be able to designate in advance the points where trains are to stop for orders, and he must do this before the necessity arises of allowing the inferior train to proceed on orders which the superior trains are subsequently to receive. If he cannot thus anticipate he must still give the order to stop for orders and send it to the point to which the meeting-order is sent, both to be delivered to the superior at the same time ; and in that case he must depend upon the signal at that point for stopping the train, as in the Standard rules, or always keep the inferior train from acting on the order until the orders for the other train are delivered.

Again, a train for which it is thought meeting-orders may have to be given must make a stop in order to get the advance order, and again another at the point named in it, perhaps only that it may receive an order annulling the first, if meeting-orders are found not to be needed. Frequently a duplicate order may be placed for a train and annulled before its arrival if the occasion for it has passed, but the advantage of this is lost if the advance order is used.

There are many roads on which the circumstances would not admit of thus always seeing far enough in advance the things to be done, and very many on which the business would not admit of the stops necessary, and the occurrence of a single exception would vitiate the whole and make it necessary to fall back on the provision "whenever practicable."

It is not easy to see how the rule could be invariably applied at junction points at which trains of superior right are to arrive from other roads or divisions, and circumstances are so various that it is difficult to determine just where such a plan could or could not be satisfactorily applied. Some say they have succeeded with it. Others point out quite conclusively that the circumstances with them are such that it would be impracticable. Where it can be applied and used without exception and the question of divided responsibility can be satisfactorily disposed of, it is, to say the least, an experiment in the right direction, but it is to be very much feared that this plan does not yet supply the universal remedy for the difficulty involved in the phrase "whenever practicable." The multiplication of messages on a busy wire will occur to all as a serious objection, but scarcely as one that should weigh against positive considerations of safety.

RULE 511.—When an order has been transmitted, preceded by the signal " 19," operators receiving it must (unless otherwise directed) repeat it back at once from the manifold copy, and in the succession in which the several offices have been addressed. Each operator repeating must observe whether the others repeat correctly. After the order has been repeated correctly, the response " complete," with the Superintendent's initials, will be given, when authorized by the Train Dispatcher. Each operator receiving this response must write on each copy the word " complete," the time, and his last name in full, and reply " i i complete " with his office signal, and will personally deliver the order to the persons addressed, without taking their signatures.

[NOTE.—On roads where it is desired, the signatures of the conductors (or conductors, enginemen, and pilots) may be taken by the operator on the delivery of the order. See also note under Rule 500. The Committee has recommended two forms of train orders—the " 31 " order and the " 19 " order ; leaving it discretionary with the roads to adopt one or both of these forms.]

This rule provides for the steps in transmission of the " 19 " order, for which signatures of trainmen are not required, as Rule 509 does for the "31" order. The steps are the same excepting as to the "O K" and its acknowledgment and the signatures. The same general considerations apply to the steps which are identical. The absence of the requirement as to signatures renders the "O K" unnecessary, the "complete" being the Dispatcher's notice both that the order has been correctly repeated and that it may be delivered after "complete" has been acknowledged, which should be in the succession in which offices are addressed. The responsibility of delivery to the right parties is placed on the operator.

The use of this method, rather than that under which trainmen sign for the order, has been the subject of much serious thought and discussion. In either case the "danger" signal and the carefulness of the operator are the means depended on for stopping a train for which an order has been transmitted. The difference is in the mode of delivery. If signatures are taken the men must take the time to go to the office. If they are not taken the men may go to the office or the operator may go out to deliver. The train may perhaps not stop entirely. In any event the delivery is likely to be hasty and without careful inspection of the order by those who receive it. A conservative view would seem to indicate that there were some risk in this, and yet many experienced officers do not look upon it in that light, and on roads having heavy traffic and many fast trains this method is used with satisfactory results.

The real solution of the question may be in careful supervision, good discipline, correct habits, and strict attention to business. In these lies *safety* ; in the opposite, *danger*.

It will be observed that a note of the Time Convention Committee, attached to the rule and here shown, indicates that the adoption of either form or both is discretionary with roads adopting the "Standard" rules, and that it is suggested that it may be provided that operators shall take the signatures of trainmen for "19" orders. These would be simply evidence of delivery, and the signatures would not, under this arrangement, be telegraphed to headquarters.

The question as to when it is best or proper to use the "19" order must be determined by circumstances. Taking and transmitting the signatures is intended to secure deliberate care in the delivery and certainty that the order is delivered to the right train.

The first is reasonably certain when the trainmen are required to go to the office and sign for the order ; the second is determined by the transmission of the signatures. Those

6

who use the "19" order must leave both these points to the care of the operator. If operators are thoroughly drilled and under constant and careful supervision, and so fully occupied with the work as to be necessarily always on the alert, this dependence is more likely to result favorably than where discipline is slack and business dull, and especially where the operator is required to attend to other duties. Circumstances may often seem to require the delivery of an order without signatures where the contrary is the usual custom. It would be necessary in such case to use special precautions in instructing the operator, and it should scarcely be allowed without special authority from the responsible head.

RULE 512.—For an order preceded by the signal "19," "complete" must be given and acknowledged for the train of superior right before it is given for the train of inferior right.

If the line fails *before an office has received and acknowledged the "complete"* to an order preceded by the signal "19," the order at that office is of no effect, and must be treated as if it had not been sent.

This rule is for the "19" order what Rule 510 is for the other, and no additional remarks are needed.

RULE 513.—The order, the "O K" and the "complete" must each, in transmitting, be preceded by "31" or "19," as the case may be, and the number of the order; thus, "31, No. 10," or "19, No. 10." In transmitting the signature of a conductor it must be preceded by "31," the number of the

order, and the train number; thus, "31, No. 10, Train No. 5."
After each transmission and response the sending operator
must give his office signal.

Here is prescribed the succession in which
the signals, etc., shall be transmitted. For the
"office signal," which the operator is required
to give after each transmission and response,
some substitute the personal signal of the operator, which is usually one or more letters assigned, by which the operator shall be known,
and indicates at the same time the operator and
the office where he is known to be on duty.

RULE 514.—The operator who receives and delivers an order must preserve the lowest copy. On this must appear the
signatures of those who sign for the order, and on it he must
record the time when he receives it; the responses; the time
when they are received; his own name; the date; and the
train number; for which places are provided in the blanks.
These copies must be sent to the Superintendent.

The subjects treated of in this rule have been
sufficiently considered in former remarks.

RULE 515.—Orders used by conductors must be sent by
them daily to the Superintendent.

This provision affords an opportunity of examining orders that have been used, and of ascertaining whether they have been prepared
and issued in accordance with the rules.

RULE 516.—Enginemen will place their orders in the clip
before them until executed.

This rule supposes that a place has been provided on each engine for placing orders con-

spicuously before the engineman who is to execute them. This is a very important provision. If he has to put them in his box or
pocket they may be rendered illegible, or for-
gotten or lost.

RULE 517.—For orders delivered at the Superintendent's
office the requirements as to record and delivery will be the
same as at other points.

This requirement would seem to be so obvious that it was hardly necessary to include it
in the rules, but for the fact that there has
been some oversight of so manifest a precaution.

RULE 518.—Orders to persons in charge of work requiring
the use of track in yards or at other points, authorizing such
use when trains are late, must be delivered in the same way
as to conductors of trains.

This rule recognizes the fact that the same
care is necessary in giving the use of the track
in the time of regular trains, whether it be to
a yard crew or a train on the road. Carelessness in this respect, by men working at stations, has frequently resulted in disaster. The
sacredness of the "rights" of trains should be
an integral part of railway doctrine.

RULE 519.—An order to be delivered to a train at a point
not a telegraph station, or while the office is closed, must be
addressed to
"*C. and E., No.* ——— (*at* ————), *care of* ———,"
and forwarded and delivered by the conductor or other person in whose care it is addressed. "Complete" will be given
upon the signature of the person by whom the order is to be

delivered, who must be supplied with copies for the conductor and engineman addressed, and a copy upon which he shall take their signatures. This copy he must deliver to the first operator accessible, who must preserve it, and at once advise the Train Dispatcher of its having been received.

Orders so delivered to a train must be compared by those receiving them with the copy held by the person delivering, and acted on as if "complete" had been given in the ordinary way.

Orders must not be sent in the manner herein provided to trains the rights of which are thereby restricted.

The subject of delivery of orders at points away from telegraph stations has already been considered. The method of doing this is here determined.

Safety in carrying this out must depend largely on the carefulness of the person selected to deliver the order.

RULE 520.—When a train is named in an order, all its sections are included, unless particular sections are specified; and each section included must have copies addressed and delivered to it.

This rule is based on the fact that all sections of a train are substantially one train, so far as schedule rights are concerned. This is definitely fixed by the "Standard" train rules. This rule provides that each section included in the operation of an order must have copies. Instances might be cited where this would seem unnecessary.

A delayed train may be ordered to meet a superior train at some point short of the meeting-point. Without any order each section of the superior train would have a right to go to the designated point, and it may be supposed that, if the first section is held by the order at that point for the inferior, the other sections cannot go by until the inferior is out of the way. While this may be true, circumstances may arise even in this case that would render it important that each section should know of the movement. The difficulty of specifying in a rule the cases in which the provision might be omitted probably led to making the rule absolute. It is pointed out, however, by practical men that serious and needless delays may often arise from strict adherence to the rule, and that in certain cases there can be no danger from giving the order to the leading section only. It is quite possible that the rule may admit of some amendment in this respect.

RULE 521.—Meeting-orders must not be sent for delivery to trains at the meeting-point if it can be avoided. When it cannot be avoided, special precautions must be taken by the Train Dispatchers and operators to insure safety.

There should be, if possible, at least one telegraph office between those at which opposing trains receive meeting-orders.

Orders should not be sent an unnecessarily long time before delivery, or to points unnecessarily distant from where they are to be executed. No orders (except those affecting the train at that point) should be delivered to a freight train at a station where it has much work, until after the work is done.

Here it is wisely provided that trains shall, if possible, be advised of their place of meeting before reaching it. It is scarcely necessary to point out the obvious reasons for this, arising from the possibility of a train, on arrival, passing the switch where the meeting is intended to be. The first and second paragraphs both suggest the advantage of being able to communicate with a train in the event of a desire to change an order or of an error having been found to have occurred on the part of a train or in the preparation or transmission of an order. The third paragraph is to guard against men forgetting orders delivered to them, through lapse of time or preoccupation in their work, and also against the necessity of changing orders issued long in advance of the time at which they are expected to be used, when a new set of circumstances may have arisen.

RULE 522.—A train, or any section of a train, must be governed strictly by the terms of orders addressed to it, and must not assume rights not conferred by such orders. In all other respects it must be governed by the train rules and time-table.

To some disciplinarians the provisions of this rule would seem to be unnecessary. To say that a thing means what it says and no more would seem to be superfluous, and yet the vital importance of the point, and the fact that it has been often disregarded, warrant this enforcement of it. A case in point came not long since to the author's knowledge. A rule in the book of a certain road required that "all trains must slow up at meeting-points with trains of any class." The rule was intended to apply to schedule meeting-points, and was so generally understood, notwithstanding the indefiniteness of the designation. An order was given requiring a superior train to wait until a time stated for the arrival of an inferior train at a point reached by the superior train before its arrival at the schedule meeting-point. The inferior train not arriving by the time stated, the superior train went on and passed the schedule meeting-point without slackening speed, as required by the rule. The inferior train was there and not quite out of the way, and a collision occurred. The conductor and engineman of the superior train claimed that the order to meet had done away with the schedule meeting-point, and therefore the rule did not apply, whereas the order was provisional, and was completely fulfilled when the inferior train failed to arrive and the superior train went on past the point named in the order without meeting the other. The inferior, being unable to reach the given point by the time stated, ran on its rights and stopped at the schedule meeting-point, respecting which the order had made no mention.

It is to be remarked that while the indefiniteness of the rule may have been partly charge-

able with the wrong view taken by the trainmen, a strict construction would make it applicable to every point that became a "meeting-
point," whether under the operation of the
rules or of special orders. A rule capable of
these different constructions is fatally defective.

RULE 523.—Orders once in effect continue so until fulfilled,
superseded, or annulled. Orders held by or issued for a
regular train which has lost its rights, as provided by Rule
107, are annulled, and other trains will be governed accordingly.

The first provision in this rule is also one
that would seem scarcely necessary, but for
the importance of emphasizing this point.
Future experience and training may render it
needless to include so simple a statement in
these rules.

Train Rule 107, referred to in the second
sentence, provides that a regular train 12
hours behind time loses all its rights, and is
practically annulled.

The expiration of orders, with the expiration, under the rules, of the entire rights of a
train which has received them, is a necessary
consequence, although to some it might not
be sufficiently clear without this authoritative
statement.

The statement that, under these circumstances, orders "are annulled," leaves the
mind in doubt as to whether they are simply

annulled by the state of facts or by the process
provided for annulling orders.   In the publi-
cation of these rules as adopted by the Pennsylvania Railroad Company this doubt is removed by modifying the language to read,
" Orders held by or issued for a regular train
are to be considered as annulled when the
train has lost its rights, as provided by Rule
No. 107, and other trains will be governed
accordingly."

The Chesapeake & Ohio road adds to Train
Rule 107 a provision that a train having the
right of track may take to a telegraph station
a train that under this rule has lost the right
to proceed.   This seems a good provision, as
such train has no right to proceed even as an
extra, and under many circumstances the Dispatcher would have difficulty in getting control of a train without this help.   The discussion of this belongs, however, more properly with the consideration of train rules.

RULE 524 (A).—A fixed signal must be used at each train-
order office, which shall display red at all times when there
is an operator on duty, except when changed to white to
allow a train to pass after getting orders, or for which there
are no orders.

When red is displayed all trains must come to a full stop,
and not proceed as long as red is displayed. The signal must
be returned to red as soon as a train has passed. It must
only be fastened at white when no operator is on duty. This
signal must also display red to hold trains running in the
same direction the required time apart. Operators must be
prepared with other signals to use promptly if the fixed sig-

nal should fail to work properly. If a signal is not displayed at a night office, trains which have not been previously notified must stop and inquire the cause, and report the facts to the superintendent from the next open telegraph office.

When a semaphore is used, the arm means red when horizontal and white when in an inclined position.

RULE 524 (B) —A fixed signal must be used at each train-order office, which shall display red when trains are to be stopped for orders. When there are no orders the signal must display white.

When an operator receives the signal "31" or "19," he must *immediately* display red, and *then* reply "red displayed." The signal must not be changed to white until the object for which red is displayed is accomplished.

While red is displayed all trains must come to a full stop, and any train thus stopped must not proceed without receiving an order addressed to such train, or a clearance card on a specified form, stating, over the operator's signature, that he has no orders for it. Operators must be prepared with other signals to use promptly if the fixed signal should fail to work properly. If a signal is not displayed at a night office, trains which have not been previously notified must stop and inquire the cause, and report the facts to the superintendent from the next open telegraph office.

When a semaphore is used, the arm means red when horizontal and white when in an inclined position.

Rules 524(A) and 524(B) refer to the character and operation of the train-order signal, and in the original report of the committee they are accompanied by a note indicating that the adoption of either or both forms of the rule is to be discretionary, according to the circumstances of traffic.

Both recognize the value of the "fixed" signal, instead of hand signals, and its necessity for the proper carrying out of the rules.

The difference between the two forms of the
rule is that the former provides that the signal shall stand constantly at "danger," excepting when changed to another position to
permit a train to pass, while with the latter
the normal position is at "safety," the other
to be shown only when an order is to be
sent.

Under the first plan a train approaching a
station must stop unless the signal is seen to
have been changed from its normal position of
"danger" to that of "safety"—from red to
white. The operator in this case moves the
signal and this is an indication that there are
no orders for that train, although there may
be for others.

The presence of an order in the hands of an
operator does not, under this method, require
that all trains passing shall stop. Under the
other plan the signal at red indicates that the
operator has orders in his hands, and no train
can be allowed to pass by the simple moving
of the signal, but each, on arrival, must stop
and get orders, or a "clearance card" stating
that there are no orders for it.

Some considerations respecting these two
methods have already been advanced, and they
need not be repeated here. There does not
seem to be any substantial reason why the
practice of permitting a train to pass, by the

movement of the signal, might not be used in connection with the plan of "normal at safety" as well as with the other, and the author is under the impression that this is done on some roads. The rule wisely requires a provision of other signals for prompt use in case the fixed signal fails to work. The machinery may break or the lights go out ; and to see that this precaution is observed is an important duty of the officer having direct supervision of these matters. The non-display of a usual night signal is recognized as a reason for inquiry and caution.

RULE 525.—Operators will promptly record and report to the Superintendent the time of the departure of all trains and the direction in which extra trains are moving. They will record the time of arrival of trains and report it when so directed.

The records and reports here required are important as a means of information for the Dispatcher and as a check on operators and trains as well as a part of the permanent record. Suitable blanks must be provided for these records.

RULE 526.—Regular trains will be designated in orders by their schedule numbers, as " No. 10" or " 2nd No. 10," adding engine numbers if desired ; extra trains by engine numbers, as "Extra 798 "; and all other numbers by figures. The direction of the movement of extras will be added when necessary, as " East " or " West." Time will be stated in figures only.

[NOTE.—In case any roads desire to state time in words as well as figures, the Committee sees no objection to their doing so.]

RULE 527.—The following signs and abbreviations may be
used:

Initials for Superintendent's signature.

Such office and other signals as are arranged by the
Superintendent.

C & E—for Conductor and Engineman.

O K—as provided in these rules.

Min—for Minutes.

Junc—for Junction.

Frt—for Freight.

No—for number.

Eng—for Engine.

Sec—for Section.

Opr—for Operator.

9—to clear the line for Train Orders, and for Operators
to ask for Train Orders.

31 or 19—for Train Order, as provided in the rules.

The usual abbreviations for the names of the months and
stations.

Rules 526 and 527 prescribe the mode of
designating trains and the use of figures, signs,
and abbreviations, with option as to figures,
in a note under Rule 526. Uniformity in these
matters is important for clearness of understanding and economy and expedition in telegraphing.

It is a question how far abbreviations may
properly be used in train telegraphing. They
certainly should be admitted only when they
can be shown not to interfere with a safe understanding of orders. Initials for the signatures
of Superintendent or Dispatcher and operators
may be used, but they would hardly be admis-
sable for the signatures of trainmen. The latter may very properly be addressed as " C. and

E." The "O K" for "all right" is an established signal, not requiring a dictionary to interpret it.

Min for minute, junc for junction, exp for express, frt for freight, eng for engine, No for number, K for o'clock, sec for section, opr for operator, cannot mislead.

For inquiries and replies respecting the work, many codes have been constructed wherein each is represented by a number or a word, and the telegraphing thus abbreviated.

It will probably never be settled to the satisfaction of everybody whether numbers should be represented in figures or written out in full. The opinion of practical men has been lately growing more favorable to figures, although some adhere rigidly to writing out numbers in words. The "Standard" rules favor figures. Much depends of course on the training of the operators. Figures are unmistakable if properly made, while a long number written out in full may be so poorly written as to confuse the reader. Where a single figure occurs in describing a section of a train as 2nd, 3rd, etc., it is easy to take the one for the other, both in telegraphing and in the written figures, and it is wise to write these out. The numbers of trains and of engines are not so liable to be confused with others in their immediate neighborhood, and it would appear to be entirely proper to use figures to represent them.

The designation of trains is usually by numbers. This is more definite and more brief than by any other time-table title, as "local freight," "Chicago express," etc. An extra train is probably best described by the engine name or number, as there is usually nothing else about a train so definite as this. Some add the names of conductors and enginemen. Where there is any danger of one train being mistaken for another, the engine number should be used, and care taken against mistakes arising from change of engines.

CHAPTER X.

The advantage of pre-arranged forms of train orders for the cases ordinarily occurring has been already adverted to, and is now fully recognized. Forms should be brief. A multitude of words is confusing. They are not so easily read ; while a short form, with a uniformly well understood meaning, is comprehended at a glance. To know what it intends becomes a part of the education of a railroad man. For this reason it would be a great advance if this service could be everywhere conducted on the same plans. Brevity also economizes time in telegraphing, which is of great importance on a busy wire. In a conversation carried on by a company of persons several may speak at once, or nearly so, and things go smoothly along, but on a wire only one can speak at a time, and hence the time each communication may occupy becomes important.

All men, however, do not quickly catch an idea when its expression is reduced to the simplest form. This is, sometimes, because it is new, or it may be from lack of training, or even natural dullness, or because human nature is so constituted that men view the sim-

7

plest things in different lights. To provide against all contingencies of this kind, and to explain to men the proper understanding as well as to settle it authoritatively, explanatory rules are needed, with definite instructions as to how orders are to be interpreted. These may be studied at leisure and discussed and mutually understood by the men. The need of these rules does not arise from any incompleteness in the forms of orders. A signal for a given purpose is sufficient in itself, but it is necessary to state the purpose which it is designed to serve. A word expresses a definite thought, but we may have to turn to the dictionary to learn what that thought is. Another and highly important service of such explanatory rules is that they beget confidence, on the ground that all understand alike.

It has been before urged that a separate order should be given for each separate transaction. This, however, need not be pressed to extremes. Circumstances may arise in which forms may be combined with advantage. For instance, an order may be given:

> *Engine* 530 *will run extra to Brighton, and will meet train No.* 2 *at Lisbon.*

This serves the purpose of an "extra" order and of a "meeting" order, and is not in any way confusing.

Ordinarily there is little to be gained by de
parting from the general rule laid down, but
experience and good judgment will soon determine where it will be proper, if the principles
upon which safety may depend are kept
steadily in view.

Attempts have been made to introduce
printed blanks for the several forms of orders,
with spaces for the words which vary with
each case, such words only to be telegraphed.
This plan does not appear, however, to have
met with much favor. The brevity possible in
forms is such that little is saved by this
method, in the amount of telegraphing. The
words sent are disconnected and unsatisfactory, and the care and attention required in
having a number of books on the operator's
table from which to select the proper form
would be considerable, especially if the manifold is used. A supposed advantage is in
having explanatory rules printed on each
blank. It is better to have these printed
together with all the forms for circulation
among the employés, who can then discuss
and become familiar with them and come to a
uniform understanding as to their meaning.

Much variety has existed in the forms of
orders in use. Prior to the quite general
adoption of the " Standard " code there were
probably no two roads on which they were in

all respects alike. This lack of uniformity was unfortunate, and some of these variations assumed serious importance in view of the time occupied in telegraphing superfluous words. A very few forms suffice for the most of the orders issued.

Those here considered are the forms issued with and forming a part of the Time Convention Rules. They are the same in principle as those given in the former edition of The Train Wire, and not greatly different in their construction. Some have been amplified and some additions have been made.

They will be considered under the following classification :

    A. For trains meeting.
    B. For trains passing.
    C. Reversing rights of trains.
    D. Movements regulated by time.
    E. For running in sections.
    F. For extra trains.
    G. For annulling trains.
    H. For annulling an order.
    I. Holding orders.

Practice may suggest additional forms or combinations of these.

In these forms trains are designated by numbers, it being understood that those of odd numbers move in one direction and have the right of track as against opposing trains of even numbers, and that the train rules fix this as well as which train shall ordinarily take the siding.

It will be understood that all orders are addressed in the manner required by the rules, including in the address the places where the order is to be delivered, thus:

C. & E. train No. 1, Paris.
C. & E. train No. 2, Madrid.

The forms are accompanied by examples of their use, with variations for different cases and explanatory notes or rules, all being a part of the "Standard" rules. Following each are the author's remarks:

Form A.—Fixing Meeting-Point for Opposing Trains.

—— and —— will meet at ——.

EXAMPLES.

No. 1 and No. 2 will meet at Bombay.
No. 3 and 2nd No. 4 will meet at Siam.
No. 5 and Extra 95 will meet at Hong Kong.
Extra 652 North and Extra 231 South will meet at Yoko-hama.

Trains receiving this order will, with respect to each other, run to the designated point, and having arrived there will pass in the manner provided by the Rules.

This order is usually given to designate a definite meeting-place at which the trains would not meet under the operation of the time-table and train rules. No. 2 has no right to pass the regular meeting-place if No. 1 is late, until it has arrived ; and No. 2. would hence in such case be delayed unless an order is given authorizing it to proceed.

If No. 2 is too late to reach the regular meeting-place before No. 1 may leave, it must, by the rules keep out of the way of No. 1 by waiting at some other point, but an order enables it to run with confidence, without time clearance, to a new meeting-place. It may happen that an order will be useful authorizing trains to meet at their regular meeting-place, when both are behind time or when the inferior train is not much late. In any case it avoids the necessity for allowing any time for clearance. It is not necessary to add to the form of the order as given above, as has been sometimes done, "and pass according to rule." The order should not be burdened with this. The rules respecting train orders should always provide, as above, that *trains ordered to meet at a designated point will both run to that point, and having arrived there will pass each other in the manner provided by the rules, unless otherwise indicated in the order* This settles the question, which has been raised, of the sufficiency of this form of order, and also renders unnecessary the expression "meet and pass." The word "pass" is best reserved for use in connection with a train going around another moving in the same direction, and it would seem unnecessary to direct trains meeting each other to "pass," as they cannot proceed without passing; and the rules should prescribe the method. This positive meeting-order is generally deemed the safest form of order for opposing trains, as it leaves no room for doubt or calculation in determining how the order is to be executed. In the use of this order for trains of several sections it must be held to apply to all the sections, unless otherwise specified, and each section that is included in the operation of the order should be referred to and is required by the "Standard" rules to have copies.

If the different sections are to be met at different places, separate orders are best. In the forms contained in a book of rules which appears to have been carefully prepared, is found the following for a train or a section of a train which is to meet one of several sections :

"Train No. — will meet and pass —— sections of train No. — as follows: first section No. —, at —— ; second section, at —— ; third section, at ——.''

Some of the objections urged against the practice of including several meeting-points in one order, under the "single order" system, apply equally to this. The whole of this order must be transcribed for and delivered to each section, and each conductor and engineman must acquaint himself with the whole, while but one train is concerned with all of it. The men of each of the sections named must carefully pick out what belongs to them, and those of the first train must exercise great care to avoid missing any of the points named. It will be found vastly better and safer to give a separate order for each meeting.

Form B. Authorizing a Train to Run Ahead of or Pass Another Train Running in the Same Direction.

(1.) ——— will pass ——— at ———.

(2.) ——— will run ahead of ———, from ——— to ———.

EXAMPLES.

(1.) *No.* 1 *will pass No.* 3 *at Khartoum.*

(2.) *No.* 4 *will run ahead of No.* 6 *from Bengal to Madras.*

When under this order a train is to pass another, both trains will run according to rule to the designated point and there arrange for the rear train to pass promptly.

Referring to Example 1, if train No. 1 is superior to No. 3, the rules should give it the right to pass, as No. 3 must keep out of its way and no order would be required. If No. 3 is the superior and is for any reason running slower than No. 1 and it is desired to permit the latter to pass, an order of this kind is needed. A regular freight train may be in the way of a special passenger train which it is necessary should pass the freight. The order may also be needed for two extras or for regular trains of equal class. If the train passed is the superior, the order does not in terms fully convey to the other all the right needed. Having passed, it may be for some time, or at a subsequent period, within the time of the superior train, and it hence would *by the train rules* be required in turn to clear the track for a train which it had passed a short time before. A fair inference is that, if allowed to pass, it is of course to proceed ahead of the other, but if this is not clearly understood or fixed by a rule, the form of the order should be modified for such cases either by adding, "and will run ahead from there," or by making it read as in Example 2 indicating the point *to* as well as that *from* which the train specified is to "run ahead" of the other.

This variation is also for authorizing a train to run ahead of and in the time of another from some point at which the other has not arrived. The point *to* which it shall so run is to be omitted when it is not desired to impose such limitation.

Under this use of the order No. 6 is assumed to be late, and No. 4, an inferior train waiting for it, is allowed to proceed in its time. No. 6 may be a first-class passenger train waiting for connections. and No. 4 may be a local freight train which is enabled by this order to proceed with its work ; or perhaps it may be a train starting from some way-station or junction at which the rules would require it to wait for No. 6 to pass. No. 6 is to assume that the other may be ahead at any point beyond that named in the order, and run accordingly. The Dispatcher of course provides, by giving more definite orders as soon as he can do so, that no unnecessary delay arises to the superior train from the operation of the order.

The train rules should make it clear that *when a train is authorized to "run ahead" of another by special order, the train following must guard against collision with the train ahead, as during the operation of the order their relative rights as to superiority (when any existed) are reversed.*

An order giving a train the right to use a given number of minutes in the time of a superior train going in the same direction, comes properly under "time-orders."

Form C.—Giving a Train of Inferior Right the Right of Track Against an Opposing Train of Superior Right.

—— has right of track against —— —— to ——.

EXAMPLES.

(1) *No. 2 has right of track against No. 1, Mecca to Mirbat.*
(2) *Extra 37 has right of track against No. 3, Natal to Ratlam.*

This order gives a train of inferior right the right of track against one of superior right to a designated point.

If the trains meet at the designated point, the train of inferior right must take the siding, unless the rules or orders otherwise indicate.

Under this order, as illustrated by example (1), if the train of superior right reaches the designated point before the other arrives, it may proceed, provided it keeps clear of the schedule time of the train of inferior right as many minutes as the inferior train was before required by the train rules to keep clear of the superior train.

If the train of superior right, before meeting, reaches a point beyond that named in the order, the conductor must stop the other train where it is met and inform it of his arrival.

Under example (2) the train of superior right cannot go beyond the designated point until the extra train arrives.

When the train of inferior right has reached the designated point, the order is fulfilled, and the train must then be governed by time-table and train-rules or further orders.

The following modification of this form of order will be applicable for giving a work train the right of track over all other trains in case of a wreck or break in the track :—

EXAMPLE.

Work Train Extra 275 has right of track over all trains between Stockholm and Edinburgh from 7 P. M. ———.

This gives the work train the exclusive right of the track between the points designated.

This form is equivalent in effect to that known as the "Regardless" order, which reads thus :

"*No. 2 will run to (Lyons) regardless of No. 1.*"

The term "regardless," although having something of a reckless sound, has been taken as exactly indicating the purport of this order, viz.: that a train is to cease to regard certain rights of another which are conferred by the rules, but are suspended or abrogated by this order. Here, as in other duplicate orders, it is understood that *a new right conferred upon one train takes away or limits a right of some other train;* and that an order allowing a train to run regardless of another requires the latter to keep out of the way.

It was thought best, and is certainly an improvement, to dispense with the old designation and adopt for this order a title and phraseology indicating its purport more specifically.

The ordinary use of this order is to advance a train to a point within the time of one superior to it, when there may be uncertainty as to the trains actually meeting there. The trains would usually proceed expecting to meet, but anticipating possible new orders. If the Dispatcher thinks he is likely to have further orders, he may find it best to add, "and ask for further orders." This will bring the trainmen at once to the office on arrival if the opposing train is not seen. A positive meeting-order is to be preferred to this form when it will as well serve the purpose. A note to this effect was proposed in the Time Convention, but it was finally determined that this should be left to the discretion of operating officers.

The use of this order for a train "running ahead," as proposed in the former edition of The Train Wire, is unnecessary with the second example under Form B.

The effect of an order in Form C is to reverse for a time or for certain parts of the track the relations of trains as respects superiority of right. Some have failed to perceive that, under certain circumstances, it will be proper for a train mentioned in this order to leave the designated point before the other has arrived.

This point is settled by the rules with the form, but it may not be altogether clear to some that the conclusion is correct. The following will perhaps make it clear:

Let A, B and C in the following diagram represent three stations, of which B is the schedule meeting-point of two trains running in the directions indicated, No. 1 being the superior train and having the right to run on its own time beyond B if No. 2 has not arrived.

A............B............C
No. 1 ☞   ☜ No. 2.

Both trains are due at B at the same time. If No. 1 is late before arriving at A an order is given:

"*No. 2 has right of track against No. 1 from B to A.*"

Under this order No. 2 becomes temporarily superior to No. 1, and obtains the right to run to A on its own time without regard to the time or rights of No. 1. On the arrival of the latter at A it may be found to have made up so much time that it can proceed toward B and reach that or some intermediate point before No. 2 can, on its own schedule time, reach such point. May it do so? There is clearly nothing in the order or in the rules to prevent. No. 1 is, for the time being, the inferior train. It is in the position of a train having no rights against No. 2, and must be governed by that fact. But any train inferior to No. 2 may go from A to B or to any point if it can clear No. 2 in accordance with the rules. It should be held as a cardinal principle in train dispatching that *an order is not to be taken as having greater effect than is actually expressed.* In the order in question one train is directed to run to a point without respect to the rights of another. This annuls the rights of the one *as respects the regular time of the other* for the portion of the track designated. The rights are simply reversed. No. 1 is now required to keep clear of the time of No. 2 as laid down in the time-table, with as much clearance as the train rules required of No. 2 as respects the time of No. 1 before the order was given. It cannot be supposed that

No. 2 may possibly run ahead of time from B. This could only be done on an order to do so duplicated to No. 1 and to any other train affected by it.

If B is the point given in the order, no such question can arise as to either train, as each is due at the same time. If, however, C is the given point, it is upon the assumption that No. 2 is too late to get farther than C without interference with No. 1. If No. 2 makes up time, so that on reaching C it is found that it has time to go farther and still keep clear of No. 1, as required by the rules, its schedule rights will admit of this, and the order does not in any way interfere with them excepting in adding to them what is supposed to be required to enable the train to reach C.

It would appear then that when an order gives a train of inferior right the right of track to a given point against a superior train, the train arriving first at the designated point may go beyond it, before the other arrives, to any point where it can clear the regular time of the opposing train the number of minutes required. The train thus passing the given point must run as the inferior of the two until the other is met, and should be required, as in the rule, to clear the other by as much as the train rules prescribe for clearance of similar trains.

As a further illustration of this question, suppose that a general order were issued giving to a regular train the right of track against all other trains. It is not to be supposed that this would prevent other trains from running, excepting as they might fall into the time of the train to which this right was given. Or the order under Form D giving all trains the right of track against a given train, does not prevent the designated train from running freely where it does not get in the way of other regular trains.

It is evident that this form of order differs from the " meeting" order in this important respect, that under certain circumstances trains may meet at some other point than that named in the order, and that it may be said that "when either train has reached the point designated in this order, it may proceed, if it can do so without trespassing on the schedule time of the other." The point is further illustrated under the operation of Form D.

It is evident that, if the inferior train is an extra, it has no schedule time by which the superior train can be guided, and hence the latter, as provided by the rule, cannot go beyond the designated point until the extra has arrived.

The careful discussion of the question here involved is justified by the fact that practical men hold different views respecting it, and many rules determine it differently or leave it wholly or partly unsettled. The fact that there is a considerable diversity of opinion upon so important a point, indicates that the course to be pursued under the circumstances should be clearly set forth in the rules. A rule should not, however, be made to add to the effect of an order. It is usually only needed by way of explanation or to authoritatively determine that upon which a doubt may exist. It may occur to some that the trains meeting at an unexpected point may not recognize each other as the trains designated in the order. It must be presumed that conductors will observe all trains met, and knowing what regular trains are due will know when they have met them, and not wait elsewhere for them ; and that extras are distinguished from regular trains by proper signals.

To avoid delays, however, a provision is made that a train of superior right reaching a point beyond that designated in the order before meeting the other train, must notify the latter when it is met. As in that case the train of superior right has not the right of track, it must take the siding where it meets the train which has been given the right of track against it. When the train of inferior right arrives at the point designated in the or-

8

der before meeting the other, the order is fulfilled ; and having no longer the right of track it must take the siding at that point or at such other point as it may reach under the operation of the rules in time to clear the train of superior right.

An order in Form C with time limit is objectionable, as there is danger of overlooking the time limit. It is better to use a distinct form for time orders.

Form D.—Giving all Regular Trains the Right of Track Over a Given Train.

All regular trains have right of track against —— between —— and ——.

EXAMPLE.

All regular trains have right of track against No. 1 between Moscow and Berlin.

This order gives to any regular train of inferior right receiving it the right of track over the train named in the order, and the latter must clear the schedule times of all regular trains, the same as if it were an extra.

This form involves the same principles as the last, and might have been included under the same general head but for the wish to give it greater distinctness. The use of "over" in the title and the rule, instead of "against" used elsewhere, is probably the result of oversight.

No form was presented by the Convention Committee for giving to a given train the right of track against all regular trains. If circumstances require, such an order can of course be given on the same plan as others involving the same principles.

Form E.—Time Orders.

(1.) —— will run —— late from —— to ——.
(2.) —— will wait at —— until —— for ——.

EXAMPLES.

(1.) *No.* 1 *will run* 20 *min. late from Joppa to Mainz.*
(2.) *No.* 1 *will wait at Muscat until* 10 *A. M. for No.* 2.

Form (1) makes the schedule time of the train named, be-tween the points mentioned, as much later as the time stated in the order, and any other train receiving the order is re-quired to run, with respect to this later time, the same as be-fore required to run, with respect to the regular schedule time. The time in the order should be such as can be easily added to the schedule time.

Under Form (2) the train of superior right must not pass the designated point before the time given, unless the other train has arrived. The train of inferior right is re-quired to run with respect to the time specified, the same as before required to run with respect to the regular schedule time of the train of superior right.

The character and effect of these two forms of Time Orders are sufficiently clear from the explanatory rules. The first simply sets back a schedule and the second is positive as to the time to which the superior train must wait. There might have been added a form author-izing an inferior train to use a given number of minutes of the time of a superior train. This would have applied to any point. The effect would have been, for the particular inferior train, the same as under Example 1 for all trains. It was probably concluded that, if a train was to run late, all others should have the benefit, and that there would be no particular advantage in a form for but one train. The time-limit feature appears also in Forms G and H.

Many object to time-orders. They are certainly not as definite as a positive meeting-order, and for this reason, and because there is a chance of error in the calculations required, they are not to be preferred. A time-table, however, is a " time order," and it is not always possible to avoid directing trains to run with reference to time. A judicious Dispatcher will discriminate as to the cases in which he should do this. In all cases such even number of minutes or hours should be given as will reduce to a minimum the risk of making the necessary addition or subtraction. The risk of a time order and of all running on time, arises largely from the possibility of trainmen not having the correct time. The allowance of five minutes for difference in watches does not appear to answer the purpose for which it is designed, as men will trespass on this. The objections made to time orders appear to be overcome as far as possible by the forms presented, and now generally adopted, with the present excellence of time-keepers and

the precautions insisted on for preserving them
in good condition.

Form F.—For Sections of Regular Trains.

—— will carry signals —— to —— for ——.

EXAMPLES.

No. 1 will carry signals Astrakhan to Cabul for Eng. 85.
2nd No. 1 will carry signals London to Dover for Eng. 90.
This may be modified as follows :
Engines 70, 85, and 90 will run as 1st, 2d and 3d sections of
No. 1 London to Dover.
For annulling a section.
Eng. 85 is annulled as second section of No. 1 from Dover.
If there are other sections following add :
Following sections will change numbers accordingly.
The character of train for which signals are carried may
be stated. Each section affected by the order must
have copies, and must arrange signals accordingly.

When two or more trains are run on the
same schedule or time-table time, with the
same schedule rights, each carrying signals for
that following it, each several train is referred
to as a "section." Upon some roads these sections following the first train are called
extra trains. This method is not recognized under the "standard" rules, the
term "extra" being applied only to trains
not run by schedule. It is of great importance that the rights of a second or other following section be clearly understood, both by
trainmen and those engaged in the issue of
telegraphic orders. The general practice is

now probably such as to leave but little misapprehension on this point, whatever may have been the case in the past, when with some the rule was to "follow the flag" wherever it might go, instead of as now treating each section, in ascertaining its rights, as though it were running alone on the schedule. When a regular train is to carry signals to denote that a second section is to follow on the same schedule, the author is of the opinion that a train order to this effect should be given in a definite form.

Rule 110 of the "Standard" rules appears to authorize the practice that prevails with some, under which the signals for freight trains running in sections are ordered on by the yard dispatcher or station agent. If the train Dispatcher is duly advised, there does not seem to be any serious objection to this, although there are reasons to be urged in favor of all orders affecting the movement of trains being issued from the central office. Certainly it would not be wise to delegate this authority as respects passenger trains, and this the "Standard" rules recognize.

The forms given for sections make the order to carry signals equivalent to an order to run as a section of a regular train. The order annulling a section implies that signals will be removed as the circumstances may require.

Form G.—For Arranging a Schedule for a Special Train.

(1.) Eng. —— will run as special —— train, leaving
—— on —— on the following schedule, and will have the
right of track over all trains:
Leave ——.
⎯⎯⎯ .
Arrive ——.

EXAMPLE.

(1.) *Eng.-77 will run as special passenger train, leaving
Turin on Thursday, Feb. 17th, on the following schedule,
and will have the right of track over all trains:*
Leave Turin 11.30 P. M.
Pekin 12.25 A. M.
Canton 1.47 A. M.
Arrive Rome 2.22 A. M.
Example (1) may be varied by specifying particular trains
over which the special shall or shall not have right of
track, and any train over which the special train is thus given
the right of track must clear its time as many minutes as
such train is required to clear the schedule time of a first-
class train.

(2.) Eng. —— will run as special —— train, leaving
—— on —— with the rights of a —— class train ——,
on the following schedule, which is a supplement to time-
table No. —— :
Leave ——.
⎯⎯⎯ .
Arrive ——.

EXAMPLE.

(2.) *Eng. 75 will run as special passenger train, leaving
Geneva, Thursday, Feb. 17th, with the rights of a first-class
train east, on the following schedule, which is a supplement
to time-table No. 10:*
Leave Geneva 10.00 A. M.
Pekin 10.30 A. M., passing No. 12.
Canton 11.00 A. M., meeting No. 7.
Arrive Athens 11.30 A. M.
Example (2) creates a regular train and the specified meet-
ing and passing points are to be regarded as if designated in
the same manner as on the time-table. Such trains will be
governed by all rules which affect regular trains.

Forms for arranging schedules were not suggested in the former edition of The Train Wire, and their use has not been very general. They appear to be adapted to some special circumstances and wants, but in the adoption of the "Standard" rules some roads have omitted a portion of the provisions under Form G.

No particular remarks need be made respecting these forms, excepting perhaps that we have here an introduction of the time feature and that any risk from this is enhanced by the considerable number of "times" to be sent by telegraph and observed by trainmen.

Form H.—Extra Trains.

—— will run extra from —— to ——.

EXAMPLE.

(a.) Eng. 99 will run extra from Berber to Gaza.

A train receiving an order to run extra is not required to guard against opposing extras, unless directed by order to do so, but must keep clear of all regular trains, as required by rule.

A "work train" is an extra, for which the above form will be used for a direct run in one direction. The authority to occupy a specified portion of the track, as an extra while working, will be given in the following form:

(b.) Eng. 292 will work as an extra from 7 A. M. until 6 P. M. between Berne and Turin.

The working limits should be as short as practicable, to be changed as the progress of the work may require. The above may be combined, thus:

(c.) Eng. 292 will run extra from Berne to Turin and work as an extra from 7 A. M. until 6 P. M. between Turin and Rome.

When an order has been given to "work" between designated points, no other extra must be authorized to run over that part of the track without provision for passing the work train.

When it is anticipated that a work train may be where it cannot be reached for meeting or passing orders, it may be directed to report for orders at a given time and place, or an order may be given that it shall clear the track for a designated extra in the following form:

(d.) *Work train 292 will keep clear of Extra 223, south, between Antwerp and Brussels after 2.10 P. M.*

In this case, extra 223 must not pass either of the points named before 2.10 P. M., at which time the work train must be out of the way between those points.

When the movement of an extra train over the working limits cannot be anticipated by these or other orders to the work train, an order must be given to such extra, to protect itself against the work train, in the following form:

(e.) *Extra 76 will protect itself against work train extra 95 between Lyons and Paris.*

This may be added to the order to run extra.

A work train when met or overtaken by an extra must allow it to pass without unnecessary detention.

When the conditions are such that it may be considered desirable to require that work trains shall at all times protect themselves while on working limits, this may be done under the following arrangements. To example (b) add the following words:

(f.) *protecting itself against all trains.*

A train receiving this order must, whether standing or moving, protect itself within the working limits (and in both directions on single track) against all trains, in the manner provided in Rule 99.

When an extra receives orders to run over working limits it must be advised that the work train is within those limits by adding to example (a) the words:

(g.) *Eng. 202 is working as an extra between Berne and Turin.*

A train receiving this order must run expecting to find the work train within the limits named.

Under Form H it has been undertaken to cover the whole subject of orders for extra trains, excepting for cases which come naturally under other forms, as when an extra is ordered to meet another train.

The term "wild" has been quite extensively used for these trains, and history should preserve the fact that on some roads, when a train was ordered to run extra, it was directed to "wildcat."

An order for a train to run extra is very simple. The train is accurately designated by the number or name of its engine, and the order reading as in example (*a*) is the foundation for those which follow.

This is of course not a duplicate order. But one train is concerned, and there is no other train to be notified until it becomes necessary to forward the extra by meeting or other orders. In those it is described as an extra and treated as any other train, but in the meantime it must keep out of the way of all regular trains, and the Dispatcher must keep it in hand and especially guard against having more than one extra on the same part of the track at the same time. Here is an element of danger where the necessities require frequent extra trains. Whenever practicable, trains should be run on a regular schedule, but it will often happen that there is no regular train upon which signals may be carried for a train that must be run, and it must go as an extra.

A precaution which has been found valuable is for the Dispatcher to have before him a large blackboard on which he shall place conspicuously the number of each extra ordered. The habit, soon acquired, of looking at this whenever an extra is ordered, has proved a sufficient safeguard where this plan has been used.

There is a class of extras which cannot be dispensed with, and the management of which gives rise to serious difficulty. These are the material or "work" trains. These trains must work upon the track away from stations, often with a large force of men, and delays to their operations cause expense as well as hindrance to work. At the same time they must not be permitted to interfere with the passage of regular trains, nor of others more than can be avoided. The solving of this problem has been attempted in various ways. Some allow the "work train" to occupy the track by right, except that it must keep out of the way of regular trains. Some permit it to work under flag "until freight trains come in sight." To get it out of the way for any but regular trains, the want must be anticipated, and an order given while it is within reach for the work train to report for orders at a designated hour and place. This plan does not give as complete control of the movements of the work train as is desirable.

A plan which has commended itself during long use, and is presented in the foregoing rules, is as follows : The work train, previous to starting out for the day, receives an order to run extra to the part of road where its work lies. At the same time, and, if convenient, in the same order, it is authorized to work upon the part of the track desired, between two contiguous telegraph stations, a specified time being added, if convenient, at which the train will have to go to one of the offices limiting the working ground, for further instructions, if it is foreseen that it may be wanted about that time for this purpose. Confining the working limits between two contiguous telegraph stations leaves the smallest practicable part of the track beyond complete control. This practically makes a section of the track for the time being a "yard," through which extras cannot pass without looking for yard engines, as is usually provided where yard rules include a portion of the main track.

The rules provide two methods for operating "work train" on the section assigned, a note by the Time Convention committee indicating that either or both may be adopted, according to circumstances. One of these requires the train to protect itself against all

trains ; the other allows it to work without
protection, and requires extras to look out
for it and protect themselves against it, after
receiving notice as to where it is working. Under the first plan the work train is
required to keep signals out at all times for its
protection, and in running to either limit of
its working ground to fully protect itself
against any extra which might come. It is of
course required to keep clear of all regular
trains, and when running to or from its working ground is provided with such meeting-
orders as may be required. Under this plan,
if the Dispatcher finds it necessary to send an
extra over the working grounds, he informs it
in the order that the work train is there (g).
This furnishes a precaution in addition to the
signals of the work train, and the proceeding
is entirely safe. It can be no less so than the
practice of working under flag in the time of a
delayed regular freight train until it appears
in sight, and this plan seems to afford an
entirely practicable method for working these
trains with the least interference with their
work and with other trains, and with entire
safety.

Under the plan by which the work train is
under no requirement to use any precautions
to protect itself on working ground, if another
extra is to pass over that ground there is only

the notice to such extra of the presence of the work train, and the necessity of protecting against it. This may be sufficient with a clear view, but there are many circumstances where the double precaution would seem to be best, as the requirement that signals shall be kept a given distance ahead of a moving train is scarcely likely to be fully complied with. The plan in which the work train is required to protect itself is not to be viewed as a case of divided responsibility, in which each party may depend on the other. The requirement for the work train is absolute. An extra getting a notice as to where the work train is employed is not required to protect itself. Such notice would lead to keeping the train under greater control and looking for the signals of the work train, and whether the rule is that the work train shall protect itself or not it would be best to give such notice, as this would enable extras to run with confidence and without protection against the work train on parts of the road where it was not employed.

As to which of the methods provided by the rule shall be used, this must depend somewhat upon circumstances. Where the passing of an extra train is very infrequent, the constant putting out of signals by the work train would seem to those charged with the duty so unnecessary that they would be likely to neglect it, and it would be better under such circumstances to require extras to protect when orders cannot be given. When extras are so frequent that the loss of time in protecting themselves would be very serious, it would be better to put the duty on the work train. There would be the advantage then of the daily habit on the part of those attending to this duty.

Form J.—Holding Order.

Hold ———.

EXAMPLES.

(1) *Hold No. 2.*

(2) *Hold all trains east.*

As any order for which " O K " has been given and acknowledged operates as a holding order for the train to which it is addressed, this form will only be used in special cases to hold trains until orders can be given or for some other emergency. The reason for holding may be added, as "for orders."

This order is not to be used for holding a train while orders are given to other trains against it which are not at the same time given to it in duplicate. It must be respected by conductors and enginemen of trains thereby directed to be held as if addressed to them. Conductors, when informed of the order, must sign for it, and their signatures must be sent and "complete" obtained.

When a train has been so held it must not go until the order to hold is annulled, or an order is given in the form :

" ——— *may go.*"

This must be addressed to the person or persons to whom the order to hold was addressed, and must be delivered in the same manner.

The rules and explanations under this form are so complete that comment as to the design and significance of the order is unnecessary. In view of much former practice, too much importance cannot be attached to the provision relating to what the holding order shall *not* be used for.

Form K.—Annulling a Schedule Train.

—— of —— is annulled.

EXAMPLES.

(1) *No. 1 of Feb. 29th is annulled.*

(2) *No. 3, due to leave Naples Saturday, Feb. 29th, is annulled.*

Adding "*from Alaska,*" or "*between Alaska and Halifax,*" when appropriate.

This order takes away all rights of the train annulled and authorizes any train or person receiving it to use the track as if the train annulled were not on the time-table.

If a train is annulled to a point named, its rights beyond that point remain unaffected.

The Train Dispatcher may direct any operator to omit repeating back an order annulling a train, until he has occasion to deliver it.

When a train has been annulled it must not be again restored under its original number by special order.

As this is a general order, which may or may not have to be delivered to trains at all telegraph stations, it is very properly provided that repeating back at once by each office need not be insisted upon.

The restoration of an annulled train under its original number would tend to confusion, and the impropriety of such action is here recognized.

When a train is annulled it naturally follows that orders previously issued to it cease to be of effect and the Dispatcher must see that the duplicates of such orders, held by other trains, are annulled, if from not doing so confusion or delay would arise. Ordinarily the order annulling the train would be sufficient, if sent to trains holding these orders. If a section of a train is annulled it would seem that the same general rule should apply. The " Standard" rules do not touch on this and it would be difficult to frame and operate a rule upon any other than the plan pointed out. It may be suggested that orders held by the annulled section should be transferred to the section following it, and which, by the rules, takes its place. This would be convenient in some cases and when so might be directed ; but there may be no following section, and, if there is, the circumstances may have so changed since the orders were issued as to render them inapplicable. The transfer of orders without the usual precautions to ensure their correct reception is objectionable and it is best to avoid it when not absolutely necessary.

The better way is no doubt to leave to the Dispatcher the disposition of orders issued for a train afterward annulled, whether such train be a section or otherwise. It would have been well if the " Standard" rules had made some explicit declaration on this point.

9

Form L.—Annulling or Superseding an Order.

Order No. —— is annulled.

This will be numbered, transmitted, and signed for as other orders.

If an order which is to be annulled has not been delivered to a train, the annulling order will be addressed to the operator, who will destroy all copies of the order annulled but his own, and write on that :

Annulled by order No. ——.

An order superseding another may be given, adding, "*this supersedes order No.* ——," or adding, "*instead of* ——."

EXAMPLE.

No. 1 and No. 2 will meet at Sparta instead of at Thebes.

An order which includes more than one specified movement must not be superseded.

An order that has been annulled or superseded must not be again restored by Special Order under its original number.

In the address of an order annulling or superseding another order, the train first named must be that to which rights were given by the order annulled or superseded, and when the order is not transmitted simultaneously to all concerned it must be sent to the point at which that train is to receive it and the required response first given, before the order is sent for other trains.

The annulling order is here properly made subject to all the safeguards adopted for orders directing the movements of trains, and placed by its number in the series with them. Superseding one order by another without the previous process of annulling is here provided for with the important provision that this method shall not be used for an order including more than one specified movement. It would seldom be applicable to such a case, and if it were it might tend to confusion, so that it is better to annul the whole order and give new instructions in separate orders.

The provision that an annulled order shall not be restored under its original number is quite necessary to avoid the confusion which might arise under the opposite course. The requirement as to priority in transmission of this order is important, in view of the fact that orders reverse the rights of trains, and the reason here is the same as that which obtains in the original transmission.

The Time Convention rules prescribe the forms, etc., for the blanks on which train orders are to be written. These forms are here shown, with the specifications for the manifold-books.

Some slight changes have been made in these by roads adopting them, but in all essential features they have not been departed from, so far as the author is aware.

Standard Train Order Blank for 19 Order.

BOUND HERE.

PERFORATED LINE.

LONDON & PARIS RAILWAY COMPANY

TELEGRAPHIC TRAIN ORDER No. —.

Superintendent's Office,   March 27, 188₅.

FORM						FORM
19	*For* STATION *to* C. & E. *of* No. 13.					19

Conductor and Engineman must each have a copy of this order.

Rec'd 2:15 P. *M. Made* Complete *at* 2:16 P. *M. Rec'd by* Jones *Op'r.*

Specifications for Train Order Form and Books for Operators for 19 Orders.

Form as here shown. Blank space for order 4 inches, with no lines. The mode of filling the blanks is indicated by small type.

Names of divisions and office to be varied to suit each division.

Form 6¾ × 6 inches below perforated line. Book 6¾ × 7½ inches.

Three hundred leaves ; stitched ; bound at top ; paper cover on face and top ; very stiff back on lower side.

Paper opaque, green, sized, and of such thickness as to admit of making 7 good copies with No. 4 Faber pencil.

To be used with carbon paper, 6¾ × 7 inches, and a stiff tin, same size, corners rounded.

Standard Train Order Blank for 31 Order.

BOUND HERE.

PERFORATED LINE.

LONDON & PARIS RAILWAY COMPANY

TELEGRAPHIC TRAIN ORDER NO.___10___

Superintendent's Office,　　　　March 27, 1885.

FORM 31	*For* STATION *to* C. & E. *of* No. 13.	FORM 31

Conductor and Engineman must each have a copy of this order.

Time received　2:15 A.　*M.*　O. K.　*given at*　2:15 A.　*M.*

Conductor.	Engine-man.	Train.	Made.	At	Received by
Jones.	Brown.	13	Complete.	2:20	Dennison.
	(Omit this column where engineman is not required to sign.)				

Specifications for Train Order Form and Books for Operators for 31 Orders.

Form as here shown. Blank space for order 4 inches, with no lines. The mode of filling the blanks is indicated by small type.

Names of divisions and office to be varied to suit each division.

Form 6¾ × 9¼ inches below perforated line. Book 6¾ × 10¼ inches.

Three hundred leaves ; stitched ; bound at top ; paper cover on face and top ; very stiff back on lower side.

Paper opaque, white, sized, and of such thickness as to admit of making 7 good copies with No. 4 Faber pencil.

To be used with carbon paper, 6¾ × 9 inches, and a stiff tin, same size, corners rounded.

The following is the clearance card proposed in connection with the "Standard" rules to be used when the train order signal is operated on the plan of Rule 524B :

LONDON & PARIS RAILWAY COMPANY
CLEARANCE CARD.

Dover,   9:15 A. M.   March 25,   188   7.

Conductor and Engineman No. __12__

I have no orders for your train. Signal is out for __No. 16.__

John Jones,

Operator.

This does not interfere with or countermand any orders you may have received.

Conductor MUST SEE that the number of HIS TRAIN is entered in the above form correctly.

Conductor and Engineman must each have a copy.

CHAPTER XI.

GENERAL REMARKS.

RULES AS TO RIGHTS OF TRACK.

The respective rights of trains are frequently spoken of in what has gone before. Any method of dispatching must be subject to modification in some of the details to accord with the particular rules of the road governing train rights. A great deal of ingenuity has been expended in constructing such rules, with a view to avoiding delay to trains under all imagined circumstances. Trains to which the superior right of track has been assigned have been required to wait at meeting points twenty, thirty or more minutes, and changing or movable rights have been connected with this, and allowances have been made for "variation in watches." These devices may occasionally prove useful, and rules are necessary to govern the trains in the most of their movements, as the telegraph may sometimes be out of order and at best cannot control the general movements of trains as well as it can be done by rule. But where the telegraph is managed with anything like the perfection now possible, the occasions are few upon which it

is unavailable for any long time ; and whatever
may have been the seeming necessity formerly
for complicated rules and time allowances, it
would seem that these may now be greatly
simplified, as has in fact been done in the
" Standard" rules.

These rules provide that all trains running
in one direction, specified on the time-table,
shall have absolute right of track over opposing trains of the same class, the rule being entirely without complication by time allowance
for clearance.

This is exceedingly simple and interposes no
difficulties in ascertaining the respective rights
of these trains. The precaution is observed of
requiring superior trains to stop at schedule
meeting-points unless the switches are seen to
be right and the track clear, and to run cau-
tiously, prepared to stop at other points where
a train may be met that has not been met at a
schedule meeting-point. This, however, adds
no complication to the rule.

For trains of different classes it is simply
arranged that those of any class shall clear the
main track five minutes before the time of those
of a superior class.

It is not within the plan of this work to enter
upon a full discussion of the various methods
of arranging train rights. It is only insisted
that the rules should be simple. This not only

tends to safety in their ordinary operation, but greatly simplifies the work of train dispatching and removes the risks to which this work is subjected by a complicated system of train rules. The reduction of the amount of mental effort required of the Dispatcher, in determining what aid he shall give to trains by special orders, reduces the risk of his making mistakes in the preparation of these orders, and the simplicity here urged is in the direct line of the work of the Time Convention committee in the preparation of the " Standard" rules.

NUMBERING SWITCHES.

Of those matters fixed by the train rules which directly affect the train dispatching, few are more important than the arrangements which determine how trains meeting shall pass each other. It is usually understood and provided that, when trains meet, those having the right of track shall keep the main track, with sometimes an exception to this in favor of trains which cannot go on the siding without backing. Where this latter provision exists it renders it unnecessary for either train to pass the switch in the face of the other when they are to meet at a siding open only at one end. It is sometimes, however, necessary to put a superior train on the siding for a train that is too heavy or too long to go on, or for some other reason. The train order must settle this, but this usually adds to its length. • The following provision has been found to entirely meet the case:

At each siding or group of switches the main track switches are numbered from No. 1, and the numbers, all running in the same general direction, are painted on the switch signals with the initial letter of the station or siding. For instance, at the London passing siding the northernmost switch will be marked L 1, and the southernmost L 2. An order is given requiring trains No. 1 and No. 2 to meet at London, and it is desired to put the superior train, No. 1, going north, on the siding. The order would then read:

No. 1 and No. 2 will meet at London No. 2.

Train No. 2 may then run to switch No. 2 on the main track, and train No. 1 can go no farther. It is a physical impossibility for the trains to pass at that switch without No. 1 going on the siding, which it would do without question under the operation of a rule requiring that *when trains meet on orders the train shall take the siding which can do so without backing.* This simple arrangement indicates also which siding is to be used at a station having several. It economizes telegraphing very much and is perfectly definite.

This plan is especially valuable when the arrangement of sidings is not of the most simple character, or when three or more trains are to meet or pass at the same point, at or near the same time. The simplicity with which the placing of the trains is effected leaves nothing to be desired. Each goes to its own place without hesitation or loss of time.

In all railroad operations we now see increased attention given to minute details. To this is due much of the marvelous advance in every department. This is especially evident in all mechanical appliances. It is very apparent in the construction of the "Standard" Rules.

The suggestion here brought forward is in this direction. Instead of directing trains to meet at a given station where there may be doubt as to the exact point, leaving them to ascertain on arrival which switch is to be used or which siding is clear, this plan gives in the order the precise point and also conveys the information as to which train will take the siding. This suggestion, made in the earlier edition of this work, has been adopted only to a very limited extent, so far as the author is aware. He is so fully convinced of its value that he feels like urging its careful consideration. To fully carry out the plan, those using the "Standard" rules would have to add the provision above indicated requiring those trains to take the siding which can do so without backing.

DOUBLE TRACK.

With more than one track the business of train dispatching is usually little more than to keep slow trains out of the way of faster ones. The protection of trains unexpectedly stopped from trains following, may be effected by the "block system" in use on many of our best roads.

Single track work may be needed when one of the tracks is blocked, but unfortunately the men engaged on double track do not become familiar with the methods for single track, and cannot usually operate them satisfactorily in emergencies.

The use of the opposite tracks for laying off trains is frequently practiced, but usually under the protection of signals only. Where there are two, three or four tracks a much more extended use of them might be made for passing trains around each other, by the adoption of the methods for single track train dispatching, with good results in the saving of sidings and in keeping heavy trains moving, and it is not improbable that expenditure for additional tracks might sometimes be postponed for considerable periods by the proper adaptation of the telegraph. There would seem to be here an opportunity for managers to keep down their capital account by increasing the capacity of their tracks by the addition of a wire. That this has not been done in many cases may have been owing to the slow advance of the science of train dispatching in past years, or perhaps to limited information on the part of railroad owners and officers as to its capabilities. It is certainly true that single track roads with siding facilities none too good are now doing an amount of business that not many years ago would have been thought to imperatively demand additional tracks.

CHAPTER XII.

Telegraphic train dispatching came with the telegraph. The first attempts were very crude. As late as the year 1865, on one of our most important railroads, the plan was for any conductor to telegraph from a station where he might be, to the conductor of an opposing train at the next station, stating when he would leave, and where he would meet the other. When the two came to an understanding they went ahead.

The early orders, in the attempt to render them more secure, were often obscured by accumulated cautions as to how to run, and by general directions. To undertake now to give the historical facts of those early days would require more research than the author has been able to give, and might involve controversy into which he does not care to enter. It appears likely that methods nearly like the present "single order" were the earliest tried, and these seem to have been more widely used than the "duplicate." The latter was at least not long behind the other. It was originated and carefully worked up in several independent quarters, and from these it has been adopted by others. The author has never used any other method. Adopting it in 1863, it was in use for some years before he was aware that others were in the same path, who may have commenced at a still earlier date.

The closing paragraph of the first edition of this work was as follows:

"This method is growing in favor, and one object of the author will have been attained if this discussion shall aid in promoting its general adoption."

In preparing this second edition the fact has constantly appeared that the former words of recommendation related to points which are now realized facts on a majority of our railroads and that the method then urged has now reached the then desired position of "general adoption."

The author cannot take leave of his subject without a special word to railroad managers. No "system" has yet been devised, or ever will be, that will work itself. Rules cannot be given to men with the expectation that they will take them up, master their principles and operate them satisfactorily, especially in so important a matter as that under discussion, without careful instruction and intelligent supervision on the part of those who, from their official position, are responsible for the results. A superintendent who is not himself particularly informed respecting the rules and methods of his telegraph department, the character and capabilities of the men employed, and the manner in which their duties are performed, cannot expect to secure the advantages which the telegraph is capable of giving. Perhaps the first public intimation that anything is wrong may be a series of so-called "accidents" on his line. Investigation points to the carelessness of some operator or dispatcher as the cause. Deeper probing would perhaps discover that such carelessness was the natural consequence of lack of constant and painstaking supervision. Besides securing for such particular supervision a competent and trustworthy person whose special business it should be, the superintendent can never get away from the necessity of constantly impressing upon such official the responsibilities of his position, discussing with him the details of the work, and seeing, at least occasionally, with his own eyes, how it is performed.

The telegraph may be viewed as holding to the railroad a relation analogous to that of the nervous system to the body. From the center of authority and intelligence it carries information and instructions to every member. It

10

keeps in motion the whole body, which, without this, would be in a measure lifeless. Its ceaseless and healthful activity is all-important; and as failure of the nervous energy is to the human frame, so to the railroad is a falling off in the vital force operating through the train wire. A tonic is needed and perhaps a change of doctors.

The author's duties for some time have not brought him into direct connection with the operation of trains, and he will probably never again be engaged in this department of railroad work.

His interest in it, however, is unabated, and his desire that the methods he has endeavored to set forth shall meet with enlarged usefulness, until better shall be found, has led him to this second effort to present what has been his study during the most of his business life, and now leads him to urge upon those now actively engaged in this work that the "price" of success, as of "liberty," is "eternal vigilance."